GCSE PASSBOOK

MATHEMATICS

Mike Ashcroft

First published 1988
Revised 1990
Reprinted 1991

Illustrations: M.L. Design

© BPP (Letts Educational) Ltd

British Library Cataloguing in Publication Data

Ashcroft, Mike
    Mathematics. — (Key facts. GCSE passbooks).
    1. Mathematics
    I. Title   II. Series
    510   QA39.2

ISBN 0 85097 808 4

Printed and bound in Great Britain by
Ashford Colour Press
Gosport, Hants.

Contents

This Passbook has been written to give help and support to students revising for the new GCSE Mathematics Examination. It covers the basic content of all syllabuses for all five Examination Groups in England and Wales (Lists 1 and 2).

It does not prepare students for extended or more difficult papers (List 3). Students wanting the most comprehensive help and advice should study the GCSE edition of *Revise Mathematics*.

The book contains many examples of GCSE-type questions. Such questions being set for GCSE are quite different from GCE 'O'-Level and CSE questions used in the past. They are, in fact, of a more 'relative to everyday happenings' type of question.

I am deeply indebted to Ray Williams (my second in the Mathematics Department at Berry Hill High School) in the writing of this book. In addition to researching and writing many of the 'Practice questions', his advice and suggestions have been most helpful. I would also like to thank my wife Norma for the excellent work she produced in the typing of the manuscript. In addition I would like to thank my editor, Michael Shardlow, and the staff of Charles Letts & Co Ltd for all their help and support throughout this project.

Introduction and guide to using the book

All mathematics courses leading to GCSE must contain a certain content which is called the CORE (List 1). Additional topics are included to complete List 2 of the syllabus but the further topics required for List 3 are not included although the contents of Lists 1 and 2 account for between 50 per cent and 70 per cent of the total marks at level 3.

The contents of the Passbook should be understood by all Mathematics students. List 1 and List 2 topics have been divided into 20 units/chapters.

You are advised to work through all of the units and attempt the revision questions at the end of each topic. In each unit you will find:

1. The aims of the unit listing what you should know or be able to do at the end of the unit
2. A clear statement of the contents of the unit with KEYWORDS clearly shown
3. 'Practice questions'. Having worked through a unit you should prepare your own summary of the main points of the unit enabling you to make a serious attempt at the 'Practice questions'.

NB: Answers are provided for the 'Practice questions' towards the end of the book.

Questions marked * are levelled at List 2 students. List 1 students should be in a position to attempt all non-starred questions and obviously List 2 students should be able to attempt all questions.

Following the answers, at the end of the book will be found some useful hints for examinations on how to approach mathematics and revision.

Throughout the book, important points to which you should pay particular attention are denoted by the special 'K' symbol.

Finally, there is an index to help in the location of specific topics within the main text.

Introduction to GCSE Mathematics

Assessment objectives are important to a student preparing for an examination. For GCSE you should be able to do the following:

1 Recall, apply and interpret mathematical knowledge in the context of everyday situations

2 Set out your work in a logical and clear manner using the correct symbols and terminology

3 Organize, interpret and present information accurately in written, tabular, graphical and diagrammatic forms

4 Perform calculations by suitable methods

5 Efficiently use an electronic calculator

6 Understand systems of measurement regularly used and use them in the solution of problems

7 Estimate, approximate and work to appropriate degrees of accuracy

8 Use mathematical and other instruments to measure and to draw to appropriate degrees of accuracy

9 Form generalizations by recognizing patterns and structures

10 Interpret, transform and make appropriate use of mathematical statements expressed in words or symbols

11 Recognize and use spatial relationships in two and three dimensions, particularly in problem solving

12 Analyse a problem, select a suitable course of action and apply an appropriate technique to obtain its solution

13 Apply combinations of mathematical skills and techniques in problem solving

14 Make logical deductions from mathematical data supplied

15 Respond to a problem of an unstructured nature by translating it into an appropriately structured form

In the case of the optional school-based component (compulsory for all students from 1991 onwards) the candidate will also be required to do some or all of the following:

16 Respond orally to mathematical questions, discuss mathematical ideas and be able to perform mental calculations

17 Carry out practical and investigational work and produce extended pieces of work

Remember, the GCSE examination demands a more positive response. You must show 'positive achievement' and get higher marks than before. This does not mean it is harder to get a Grade C, for example, or that the papers are different.

Unlike GCE and CSE examinations, you are not competing with other candidates. If you can show the examiner that you can meet the assessment objectives, the grade you achieve will reflect how well you have met them.

Hints on how to approach Mathematics examinations

The most important part of the examination preparation is to make sure that you have done adequate revision. In order to do this you should consider how best to organize your revision in view of the fact that you have other subjects to consider. Time spent in considering the organization and method of your revision can reap good rewards.

You should plan to begin your revision at least six weeks before any exam and you should ask yourself the question: 'When, where and how shall I revise?' How you allocate your time to each subject will depend to a large extent on the demands of each subject and also on the make-up of the exam timetable.

Probably the most suitable time to study Mathematics, which demands much concentration, is the early evening before your brain becomes tired. It is also important to note that while hard work is essential, it is important not to overstretch yourself but to be sensible and allow yourself some leisure time.

In answer to the question of where to revise, you really need a quiet, peaceful room in which to work. Television and background music do not provide a good working environment. Sitting at a desk or table is by far the best position in which to revise Mathematics, because it is based on pencil and paper work rather than simply reading books.

How to revise? This is a very complicated question to answer since everyone is different. However, you will find some of the following points of help to you.

Obtain a copy of your syllabus and split the work into small amounts. Set yourself targets for a session or group of sessions. For each session have an aim and a plan of what you wish to achieve in that session. Allow yourself a set time in which to achieve this aim and take a break at the end of that time.

Remember that Mathematics can only be properly revised by actually 'doing them' – i.e. solving actual problems, whether they are examples from worksheets, past papers or specimen questions. Besides being a sensible way to revise, this is also a good opportunity to learn exam technique. You become familiar with the different styles of questions and appreciate the necessity to read them carefully in order to understand them.

Always write your answers in full as you would in an exam, clearly and precisely, and then ask yourself the questions: 'Can the examiner follow my argument? Have I been accurate and included all the relevant stages in my solution?'

If you have difficulty with a particular question then use your notes or text-book and make a summary of the 'key facts' in order to help you solve the problem. This will also help you to retain the material for use with further problems of a similar type.

If the text-notes do not solve your difficulties try a few of the following points:

1 Reread the question to see if you have misread it.

2 Try to remember if you have encountered a similar problem previously.

3 Attempt to break down a problem into smaller parts, each of which may be tackled individually. Some questions, although lengthy in full, consist of a number of small and quite straightforward parts.

4 Whenever possible use a diagram to help you to see the question better.

5 Leave the problem for a while, have a break and return to it later. You can then make a completely fresh start on it.

6 If you still cannot sort out the problem, note your difficulties and consult your teacher as soon as possible.

Finally, make sure you know which formulae will be provided for you in the syllabus and exam and learn those which are not. If you are in any doubt at all concerning your exam, ask your teacher for advice – he/she will be only too happy to sort problems out for you, particularly at this stage.

Sample revision programme

You should prepare your own revision programme. Let us assume you start about 6 or 7 weeks before the examination.

Days 1–20 Spend one day on each unit. Write a brief summary in your own words to use later.

Days 21–25 Choose the five units you found most difficult. Look at them again and ask your teacher for help.

Days 26–40 Work through each chapter again. Look for questions from books and past examination papers to try.

Days 41–45 Look through your summary sheets and check back on anything which is not clear. Make a special effort to learn the 'key facts' marked in this book.

Aims of the unit

To revise the various number systems and definitions that we need to be familiar with.

Integers

These are negative and positive whole numbers. But 0 (zero) is neither positive nor negative – it is just an integer.

$$\{..., -4, -3, -2, -1, 0, +1, +2, +3, ...\}$$

Natural numbers

$$\{1, 2, 3, 4, ...\}$$

These are numbers which are positive integers with 0 (zero) excluded.

Rational numbers

$$\{\tfrac{1}{3}, \tfrac{1}{2}, \tfrac{3}{4}, 0\cdot8, \tfrac{4}{1}, \tfrac{19}{2}, ...\}$$

These are numbers which can be expressed as a fraction or a ratio.

$0\cdot8 = \tfrac{8}{10} = \tfrac{4}{5}; \ \tfrac{4}{1} = 4.$

Irrational numbers

$$\{\pi, \sqrt{2}, \sqrt{3}, \sqrt{5}, \sqrt{6}, ...\}$$

These are numbers which cannot be expressed as fractions or ratios. For instance, $\pi = 3\cdot1415926$ ... to seven decimal places; it has been calculated to thousands of decimal places, but the decimal has never ended.

Odd numbers

$$\{1, 3, 5, 7, ...\}$$

Numbers which will not divide by 2 exactly.

Even numbers

$$\{2, 4, 6, 8, ...\}$$

Numbers which will divide by 2 exactly.

Prime numbers

$$\{2, 3, 5, 7, 11, ...\}$$

A prime number is a whole number which can only be divided by itself and 1. 0 and 1 are not considered to be prime numbers.

Square numbers

These are formed by multiplying a whole number by itself.

Example: $5 \times 5 = 25$, so 25 is a square number. We often write 5×5 as 5^2 ('five squared'). Hence the set of square numbers will be $\{1^2, 2^2, 3^2, ...\}$, i.e. $\{1, 4, 9, ...\}$. In diagram form, square numbers can be represented by dots in the form of a square as in Fig. 1.

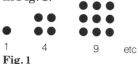

1   4   9   etc

Fig. 1

Cubic numbers

$$\{1^3, 2^3, 3^3, 4^3, ...\}, \text{i.e.} \ \{1, 8, 27, 64, ...\}$$

Triangular numbers

$$\{1, 3, 6, 10. \ ...\}$$

These numbers can be arranged in dot form to show how **equilateral triangles** are formed, Fig. 2.

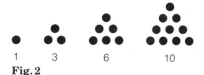

1   3   6   10

Fig. 2

Square root

This is the reverse operation of squaring (above), the symbol used is $\sqrt{\ }$ and hence $\sqrt{25} = 5$, $\sqrt{100} = 10$, etc.

Factors

If a number divides exactly into another number (i.e. without leaving a remainder) the first number is called a FACTOR of the second. Example: 2 is a factor of 8 (since $8 \div 2 = 4$, a whole number).

Multiples

If a number divides exactly into another number, the second number is called a MULTIPLE of the first. Example: 8 is a multiple of 2; 15 is a multiple of 3; etc.

Highest common factor (HCF)

The HCF of two or more numbers is the largest number which will divide into them exactly (i.e. without leaving a remainder). Example: the HCF of 72 and 96

$$72 = 3 \times 24, \quad 96 = 4 \times 24$$

Hence 24 is the HCF of 72 and 96.

Indices

When we multiply a row of numbers which are all the same value, we get a POWER of the number. Example: $9 \times 9 \times 9 \times 9 \times 9$ is the fifth power of 9 and in index form is written as 9^5.

If the index is negative, i.e. a^{-m} then this can be rewritten as $\dfrac{1}{a^m}$,

e.g. $4^{-2} = \dfrac{1}{4^2} = \dfrac{1}{16}$

If the index is fractional, i.e. $a^{\frac{1}{m}}$ then this can be rewritten as $^m\sqrt{a}$.
For example, $49^{\frac{1}{2}} = \sqrt{49} = 7$ or $125^{\frac{1}{3}} = {}^3\sqrt{125} = 5$.

Lowest common multiple (LCM)

The LCM of two or more numbers is the smallest number which is exactly divisible by them. Example: the LCM of 4, 8, 20

Factors of $\ 4 = 2 \times 2\ (2^2)$
Factors of $\ 8 = 2 \times 2 \times 2\ (2^3)$
Factors of $20 = 2 \times 2 \times 5\ (2^2 \times 5)$

Hence the LCM of 4, 8, 20 $= 2^3 \times 5 = 40$.

Order of performing operations

We must always observe a particular sequence when we are faced with a jumble of operations. The order is as follows:

1 brackets
2 multiplication/division
3 addition/subtraction

Examples $\ 5 + 12 \div 4 = 5 + 3 = 8$
$\qquad\qquad 19 - (7 + 3) = 19 - 10 = 9$
$\qquad\qquad (6 + 8) \div (10 - 8) = 14 \div 2 = 7$

Sum of numbers

This is the added total.

Difference of numbers

The answer when one number is subtracted from another.

Product of numbers

The result of multiplying numbers together.

Quotient of numbers

The result of dividing one number by another.

Symbols

$=$ is equal to $\neq$ is not equal to
$>$ is greater than $<$ is less than
$\geq$ is greater than or equal to $\leq$ is less than or equal to

Practice – Unit 1   Number

1 Write the next two numbers in the series:
(a) 1, 4, 9, 16, ..., ...
(b) 1, 3, 6, 10, 15, ..., ...
(c) 1, 2, 3, 5, 8, 13, ..., ...
(d) 1, 8, 27, 64, ..., ...
2 Work out the value of the following:
(a) $\sqrt{144}$. (b) $\sqrt{64} \div \sqrt{4}$. (c) $\sqrt{25 \times 4}$. (d) 8^3. (e) 13^2. (f) 2^5.
3 Examine this set of numbers: {7, 15, 16, 24, 50}.
From the set write down:
(a) A square number.
(b) A prime number.
(c) A multiple of 12.
(d) The product of the odd numbers.
(e) The sum of the even numbers
(f) A factor of 60.
4 Work out the value of these:
(a) $17 - (8 - 3)$.   (b) $(17 - 8) - 3$.
(c) $(4 \times 6) + 2$.   (d) $4 \times (6 + 2)$.
(e) $(24 \div 3) + 5$.   (f) $24 \div (3 + 5)$.
5 Table 1 gives information about the attendance at four football grounds at four consecutive home matches. Fill in the spaces.

Table 1

	Arsenal	Man. Utd	Leeds	Orient	Total
Week 1	21 461	39 586	31 429	9628	102 104
Week 2	24 072	42 721	27 333	5406	?
Week 3	?	41 027	29 281	9672	?
Week 4	20 467	49 678	?	8241	105 007
Club Total	93 624	?	?	32 947	?

6 (a) What are the common factors of 15 and 20?
(b) What is the highest common factor of 24 and 60?
(c) What is the lowest common multiple of 8 and 12?
(d) What is the lowest common multiple of 3, 4 and 8?
(e) Bell A is rung every 5 seconds. Bell B is rung every 8 seconds. Bell C is rung every 12 seconds. If they are started together, after how many seconds do they ring together for a second time?

7 Calculate the value of the following, giving your answers as fractions where necessary:

(a) 5^{-1} (b) 2^{-1} (c) 4^{-2} (d) 2^{-3} (e) 5^0

(f) $4^{\frac{1}{2}}$ (g) $25^{\frac{1}{2}}$ (h) $8^{\frac{1}{3}}$ (i) $27^{\frac{1}{3}}$ (j) $100^{\frac{1}{2}}$

Unit 2 Fractions, decimals, and percentages

Aims of the unit

To revise:
1 Vulgar fractions (language and notation),
2 The four rules applied to vulgar and mixed fractions,
3 Decimal fractions and the four rules,
4 Percentages, percentage change,
5 Conversion between vulgar, decimal fractions and percentages.

Equivalent fractions

In Fig. 1 we can see that $\frac{3}{4} = \frac{6}{8}$ and these are known as EQUIVALENT fractions. $\frac{3}{4}$ is in fact $\frac{6}{8}$ in its lowest terms.

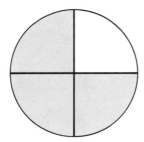 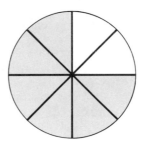

Fig. 1

Other fractions equivalent to $\frac{3}{4}$ include $\frac{9}{12}, \frac{12}{16}, \frac{15}{20}$, etc.

Vulgar fractions

If the numerator of a fraction is smaller than the denominator, the fraction is called a VULGAR or proper fraction.
Example $\frac{2}{3}$ is a vulgar fraction.

Improper fractions

If the numerator of a fraction is larger than the denominator, the fraction is called IMPROPER.
Example $\frac{8}{5}$ is an improper fraction.

Mixed number

If a number consists partly of an integer and partly of a fraction, it is called a MIXED NUMBER.

Example $2\frac{4}{5} = 2 + \frac{4}{5}$ (i.e. a mixed number).

To change the improper fraction $\frac{19}{8}$ to a mixed number, divide 19 by 8, giving an answer of 2 rem. 3. This represents 2 whole numbers and the remainder of 3 will be $\frac{3}{8}$. Hence the mixed number is $2\frac{3}{8}$.

To change the mixed number $3\frac{2}{7}$ to an improper fraction, multiply the 3 whole numbers by 7 giving 21 sevenths and then add the 2 sevenths giving a total of $\frac{23}{7}$.

LCM (least common multiple)

Looking at the fractions $\frac{2}{3}, \frac{7}{12}, \frac{3}{4}$ we can try to put them in order from the largest down to the smallest. In order to do this, we must find a common denominator which is suitable for all the denominators, i.e. 3, 12, 4. Obviously 12 is the least common denominator (or multiple) and we then convert all the fractions into twelfths.

$$\frac{2}{3} = \frac{2}{3} \times \frac{4}{4} = \frac{8}{12}, \quad \frac{7}{12} = \frac{7}{12}, \quad \frac{3}{4} \times \frac{3}{3} = \frac{9}{12}$$

Hence $\frac{3}{4}$ is the largest and $\frac{7}{12}$ is the smallest and so $\frac{3}{4}, \frac{2}{3}, \frac{7}{12}$ would be the correct order (largest to smallest).

Adding/subtracting fractions

Only fractions having the same denominators may be added/subtracted and so if the denominators are not the same then they must be made so by finding their LCM.

Also note that if mixed numbers are involved then the integers should be added/subtracted first of all.

Example 1 $2\frac{3}{4} + 4\frac{7}{12} = 6 + \frac{9}{12} + \frac{7}{12} = 6\frac{16}{12} = 7\frac{4}{12} = 7\frac{1}{3}$ (in lowest terms).

Example 2 $4\frac{7}{8} - 2\frac{5}{6} = 2 + \frac{21}{24} - \frac{20}{24} = 2\frac{1}{24}$ (LCM = 24).

Example 3 $3\frac{2}{5} - 1\frac{3}{4} = 2 + \frac{8}{20} - \frac{15}{20} = 2\frac{8-15}{20}$. In order to overcome the problem of $8 - 15$ we must use one of the 2 whole numbers and convert it into $\frac{20}{20}$ which is then added to the $\frac{8}{20}$ to give $\frac{28}{20}$.

The problem can then be rewritten as $1\frac{28-15}{20} = 1\frac{13}{20}$.

Multiplying fractions

To multiply a fraction by a fraction (or whole number) we must multiply the numerators together to give the numerator of the answer and then multiply the denominators together to give the denominator of the answer.

Example 1 $\frac{2}{3} \times \frac{4}{5} = \frac{8}{15}$.

If we have mixed numbers to be multiplied, then they must be converted to improper fractions before the multiplications are done.

Example 2 $5\frac{1}{3} \times 2\frac{1}{4} = \frac{16}{3} \times \frac{9}{4} = \frac{144}{12} = \frac{12}{1} = 12.$

Dividing fractions

To understand this process, we must realize that dividing by 2 is the same as multiplying by $\frac{1}{2}$. Similarly dividing by $\frac{4}{5}$ is the same as multiplying by $\frac{5}{4}$.

Using this idea, we can change any division problem into a multiplication problem and then proceed as shown above. Once again, before starting, any mixed numbers must be converted into improper fractions.

Example 1 $6\frac{3}{7} \div 5 = \frac{45}{7} \div \frac{5}{1} = \frac{45}{7} \times \frac{1}{5} = \frac{45}{35} = \frac{9}{7} = 1\frac{2}{7}.$

Example 2 $2\frac{5}{8} \div 8\frac{1}{6} = \frac{21}{8} \div \frac{49}{6} = \frac{21}{8} \times \frac{6}{49} = \frac{126}{392} = \frac{18}{56} = \frac{9}{28}.$

Fractional amounts

We must be able to calculate fractional parts of certain amounts of money, length etc.

Example   What is $\frac{3}{4}$ of £50? (NB 'of' means *multiply*.)

     Now, $\frac{1}{4}$ of £50 = £12.50

     and so $\frac{3}{4}$ of £50 = $3 \times$ £12.50 = £37.50

Practice – Unit 2   Fractions

1   What fractions are shaded in these:

(a)   (b)   (c)

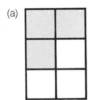

Fig. 2

2   Shade in the fraction stated:

(a)   (b)   (c)

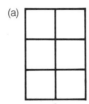
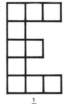
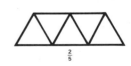

Fig. 3 $\frac{2}{3}$ $\frac{1}{5}$

3 Work out the answers to the following, writing your answers as fractions in their lowest form:

(a) $2\frac{3}{4} - 1\frac{1}{2}$. (b) $\frac{3}{5} - \frac{1}{2}$. (c) $2\frac{1}{4} \times 5$. (d) $\frac{1}{4} \times \frac{2}{5}$.

(e) $2\frac{1}{2} \div 5$. (f) $3\frac{1}{3} \div 1\frac{1}{2}$. (g) $10 \div 2\frac{1}{2}$. (h) $\frac{1}{2} + 2\frac{1}{4} + 1\frac{3}{4}$.

4 A man wins £7500 on the pools and decides to share it with his family. He gives his wife $\frac{2}{5}$, his son $\frac{1}{4}$ and saves the rest for himself.
(a) How much does he give to his wife?
(b) How much does he give to his son?
(c) What fraction does he save for himself?

*5 A swimming pool is filled at a constant rate. When half full the pool has had 7500 litres poured into it.
(a) How much will the pool hold when full?
(b) What fraction has been filled when 5000 litres of water are in the pool?
(c) What fraction has still to be put into the pool?
(d) How much water will be needed when 5000 litres is already in the pool?

*6 Pete spends half his wages on living expenses and saves a quarter of what he has left. He gives his mother all that remains which is £36.
(a) What did he save?
(b) What did he earn?
(c) What fraction did he give to his mother?

*7

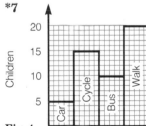

Fig. 4

In a survey of children travelling to school the results were as shown in Fig. 4.
(a) How many children took part in the survey?
(b) What fraction of the survey walked to school?
(c) What fraction travelled by bus?
(d) What fraction cycled to school?

Decimal fractions

We must be absolutely sure that we understand the meaning of each digit in a decimal number. The purpose of the decimal point is to separate the whole numbers from the fractional parts, i.e. 15·6 represents 15 whole ones and 6 tenths. 8·658 represents 8 whole ones and 658 thousandths or 6 tenths, 5 hundredths and 8 thousandths.

Adding/subtracting decimals

Remember that, when adding and subtracting in decimal form, it is helpful if we work in vertical columns and the decimal points must be under one another. The addition/subtraction is then carried out in exactly the same way as for the ordinary number system.

Example $0 \cdot 328 + 7 \cdot 63 + 6$. (NB '6' means 6 whole ones, i.e. $6 \cdot 0$.)

Hence:
$$
\begin{array}{r}
0 \cdot 328 \\
7 \cdot 630 \\
6 \cdot 000 \\
\hline
13 \cdot 958
\end{array}
$$
← (noughts are added without changing the problem)

Multiplying decimals

To carry out this operation we ignore the decimal points and multiply as for the ordinary number system, placing the decimal point in its correct position afterwards.

Example $14 \cdot 7 \times 0 \cdot 59$.

$$
\begin{array}{r}
147 \\
\times 59 \\
\hline
1323 \\
7350 \\
\hline
8673
\end{array}
$$

We can say that the answer will have as many decimal places (numbers after the decimal point) as the number of decimal places in the two numbers we are multiplying, added together.

In the above case $1 + 2 = 3$ decimal places required. Hence $14 \cdot 7 \times 0 \cdot 59 = 8 \cdot 673$.

Dividing decimals

In order to do this operation we must always have the DIVISOR (the number we are dividing by) as a whole number.

Example $17 \cdot 04 \div 0 \cdot 8$. In this case we need the divisor as 8, not $0 \cdot 8$. To do this we multiply $0 \cdot 8$ by 10; but also we have to multiply $17 \cdot 04$ by 10 giving $170 \cdot 4$. The problem now becomes $170 \cdot 4 \div 8$ which can be dealt with as for ordinary short division, i.e.

$$
\begin{array}{r}
21 \cdot 3 \\
8 \overline{)170 \cdot 4}
\end{array}
$$

Hence $17 \cdot 04 \div 0 \cdot 8 = 21 \cdot 3$.

Changing fractions to decimals

To change a fraction into a decimal we divide the NUMERATOR by the DENOMINATOR.

Example $\frac{3}{4} = 0.75$; $\frac{4}{5} = 0.8$.

Even the most awkward vulgar fractions can be converted using a calculator to assist.

Example $\frac{21}{25} = 0.84$.

Changing decimals to fractions

To change a decimal into a fraction, we must see how many decimal places are involved and act accordingly.

Example 1 0.5 (one decimal place – denominator 10). Then $0.5 = \frac{5}{10} = \frac{1}{2}$ (in lowest terms).

Example 2 0.625 (three decimal places – denominator 1000). Then $0.625 = \frac{625}{1000} = \frac{5}{8}$ (in lowest terms).

Standard form

A number expressed in the form $A \times 10^n$, where A is a number between 1 and 10 and n is a positive or negative integer, is written in STANDARD FORM. It is particularly useful for expressing very large or small numbers.

Example 1 $5\,300\,000\,000 = 5.3 \times 10^9$ in standard form. To obtain this:

1 Place the decimal point between the first and second numbers so that $A = 5.3$.
2 Count the number of decimal place moves to the *right* required to restore the decimal point to its original position. It is 9. Therefore 5.3 is multiplied by 10^9. Hence the number $= 5.3 \times 10^9$.

Example 2 The very small number $0.000\,000\,054 = 5.4 \times 10^{-8}$. To obtain this:

1 Proceed as before to make $A = 5.4$.
2 To restore the decimal point to its original position it must be moved 8 places to the *left*. 5.4 is divided by 10^8 or multiplied by 10^{-8}. Hence the number $= 5.4 \times 10^{-8}$.

NB: In standard form, numbers less than one always have a negative power of 10.

Practice – Unit 2   Decimals

1   Write, in order of size, smallest first: $1\cdot65$; $1\frac{1}{2}$; $1\frac{2}{3}$; $1\cdot66$; $1\frac{3}{5}$.

2   Use your calculator to complete the following:

(a) $\dfrac{4\cdot8 \times 2\cdot4}{1\cdot6} = \ldots.$ (b) $\dfrac{4\cdot8}{2\cdot4 \times 1\cdot6} = \ldots.$

(c) $1\cdot5^2 - 0\cdot9^2 = \ldots.$

3   **Table 1**

1 litre = 1·8 pints
1 inch = 2·5 cm

Use Table 1 to complete:
(a)   3 litres   = ... pints
(b)   15 inches = ... cm
(c)   15 cm   = ... inches

4   Mr Jones's car averages 38·5 miles to the gallon. Petrol costs £1.80 per gallon.
(a)   How far will Mr Jones travel on 6 gallons?
(b)   What will be the cost of petrol for this journey?
(c)   How many gallons will he need to travel 308 miles?
(d)   If his petrol bill comes to £19.80, how many gallons did he put in?

5   The four runners in a 4×400 metres relay race record times of: 45·64 secs; 46·21 secs; 47·14 secs; 48·25 secs.
(a)   What is their total time in seconds for the whole race?
(b)   What is their total time in minutes and seconds for the whole race?
(c)   What is the time gap between the fastest and slowest runner?

*6   Write these numbers in standard form:
(a)   6700   (c)   0·074
(b)   845 000   (d)   0·000 38.

*7   Write these as ordinary numbers:
(a) $1\cdot7 \times 10^3$ (c) $2\cdot3 \times 10^1$
(b) $1\cdot04 \times 10^5$ (d) $6\cdot2 \times 10^{-1}$.

Percentages

PERCENTAGE means 'part of a hundred'. We usually use percentages when we talk about examination marks at school or about money involving discounts, profits etc. For instance, 50 per cent (50%) means 50 parts of a hundred parts, which can be written as a fraction $\dfrac{50}{100}$ or $\dfrac{1}{2}$ (in lowest terms).

Changing percentages to fractions/decimals

Percentages are simply fractions with a denominator of 100.

Example: $20\% = \dfrac{20}{100} = \dfrac{1}{5}$ in lowest terms $= 0.2$ in decimal form.

Changing fractions/decimals to percentages

In order to do this we simply multiply the fraction/decimal by 100.

Example 1 $\dfrac{4}{5} = \dfrac{4}{5} \times 100\% = \dfrac{400\%}{5} = 80\%.$

Example 2 $0.65 = 0.65 \times 100\% = 65\%.$

Finding a percentage of something

Example 1 Find 10% of £80 $= \dfrac{10}{100} \times £80 = \dfrac{800}{100} = £8.$

Example 2 Find 5% of £60 $= \dfrac{5}{100} \times £60 = \dfrac{300}{100} = £3.$

Changing marks into percentage marks

Example 1 30 out of 50 $= \dfrac{30}{50} \times 100\% = \dfrac{3000}{50} = 60\%.$

Example 2 45 out of 60 $= \dfrac{45}{60} \times 100\% = \dfrac{4500}{60} = 75\%.$

Increasing/decreasing by a percentage amount

Example 1 Increase £30 by 10%.

     10% of £30 = £3

therefore increasing £30 by 10% gives £30 + £3 = £33.

Example 2 Decrease £40 by 5%.

     5% of £40 = £2

therefore decreasing £40 by 5% gives £40 − £2 = £38.

Percentage increase/decrease (change)

This is found by:

$$\dfrac{\text{Increase/Decrease}}{\text{Original}} \times 100\%$$

Example The population of a town in 1960 was 16 000 and in 1980 was 20 000; what was the percentage increase in population?

     Increase in population = 4000

$$\% \text{ increase} = \dfrac{\text{Increase in population}}{\text{Original population}} \times 100\%$$

$$= \frac{4000}{16\,000} \times 100 = \frac{400\,000}{16\,000} = 25\% \text{ increase}$$

Practice – Unit 2   Percentages

1 Complete Table 2, linking corresponding fractions, decimals and percentages:

Table 2

Fraction	Decimal	Percentage
$\frac{1}{2}$	0·5	50%
$\frac{1}{4}$		
	0·75	
		30%
$\frac{2}{5}$		
	0·8	
		60%
$\frac{7}{25}$		
	0·15	
$\frac{11}{20}$		
	0·666 66	
$\frac{1}{3}$		
		$12\frac{1}{2}\%$

2 Peter's results in his end-of-term tests were as follows:

Maths:	39 out of 50	English:	13 out of 20
Science:	15 out of 25	History:	7 out of 10
Geography:	24 out of 40		

Write each result as a percentage mark.
Place the results in order, with his best mark first.

3 In a class of 30 pupils, 40% are boys.
(a) What percentage of the class are girls?
(b) How many boys are there in the class?

4 A shopkeeper buys some watches at £25 each. He tries to sell them at a 40% profit.
(a) What price does he advertise them at?
In the end he puts them in a sale at 10% off the advertised price.
(b) How much does he sell them for?
(c) What is his profit per watch?
(d) Write this profit as a percentage of the initial cost price.

*5 (a) Mrs Jones's wages are £82, but she gets a 5% pay rise. What is her new wage?
(b) Mrs Smith earns £84 after a 5% pay rise. What was her wage before the pay rise?

*6 A motorbike is sold for £720 at a loss of 20% on its original purchase price. What was the original price?

*7 (a) Last year I was 160 cm tall. Now my height has increased to 168 cm. What was my percentage increase in height?
(b) This year my sister is 159 cm tall. During the previous year her height had increased by 6%. How tall was she last year?

Unit 3 Directed number

Aims of the unit

To revise:
1 The number line,
2 Use of greater/less than symbols,
3 Directed number operations – four rules,
4 Simple problems involving directed number.

The number line

We can use either a vertical or a horizontal line (Fig. 1) to show how negative and positive numbers may be linked.

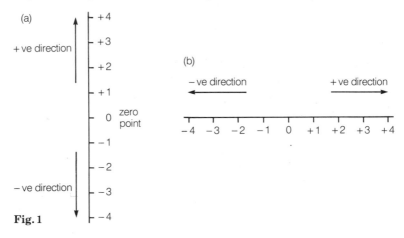

Fig. 1

We can see that the extended number line (Fig. 2) travels in two opposite directions. The numbers to the right of zero are positive. They get bigger and bigger as we go along the number line to the right. The numbers to the left of zero are negative. They get smaller and smaller as we go along the number line to the left.

Fig. 2

'Greater than' and 'less than'

       $9 > 3$ means that 9 is greater than 3.
       $4 < 6$ means that 4 is less than 6.

If we apply the signs to the number line, then we can see that $+3 > -4$ (positive 3 is greater than negative 4).

The statement could be reversed to read: $-4 < +3$ (i.e. negative 4 is less than positive 3).

Other examples: $0 > -1$, $-2 > -4$, $-5 < 0$ and $-9 < -8$.

Adding directed numbers

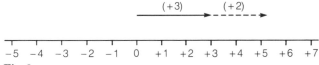

Fig. 3

Example 1 $(+3) + (+2) = +5$ (see Fig. 3).
Example 2 $(-3) + (+5) = +2$.
Example 3 $(+7) + (-3) = +4$.
Example 4 $(-2) + (-3) = -5$.

Subtracting directed numbers

Subtraction is finding the difference between two amounts.

Example 1 The difference between $+6$ and $+2$ can be $+4$ or -4 depending on which way round we compare the two directed numbers.

Fig. 4

Compared with $+2$, $+6$ is a move of 4 to the right, hence, $(+6) - (+2) = +4$ (Fig. 4).

However, compared with $+6$, $+2$ is a move of 4 to the left, so $(+2) - (+6) = -4$ (Fig. 5).

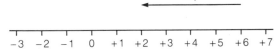

Fig. 5

Example 2 Consider $(+3) - (-2)$.

Using the number line in Fig. 6, compared with -2, $+3$ is a move of 5 to the right, so $(+3) - (-2) = +5$.

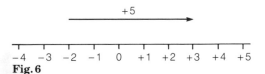

Fig. 6

With a little practice it becomes quite easy to solve the problems without using the number line. Consider the following:

(a) $(+6)-(+3) = +3.$ (b) $(+4)-(-3) = +7.$
(c) $(-4)-(-5) = +1.$ (d) $(-5)-(+3) = -8.$
(e) $(+1)-(+5) = -4.$

Multiplying and dividing directed numbers

K▶ Under the operation of multiplication, the directed numbers behave as shown in the following examples.

Example 1 Two positive numbers multiplied together give a positive product, i.e.
$$(+3)\times(+4) = +12$$

Example 2 A positive number and negative number multiplied together give a negative product, i.e.
$$(+4)\times(-5) = -20$$

Example 3 Two negative numbers multiplied together give a positive product, i.e.
$$(-2)\times(-5) = +10$$

Division of directed numbers follows the same pattern as multiplication, i.e.
$$(+32)\div(+8) = +4$$
$$(+24)\div(-8) = -3$$
$$(-36)\div(-3) = +12 \text{ etc.}$$

Hence, $(+) \overset{\div}{\underset{\times}{}} (+) = +\text{ve result}$

$(+) \overset{\div}{\underset{\times}{}} (-) = -\text{ve result}$

$(-) \overset{\div}{\underset{\times}{}} (+) = -\text{ve result}$

$(-) \overset{\div}{\underset{\times}{}} (-) = +\text{ve result}$

Directed number in practical situations

Example The temperature in Moscow was $-9°C$ when it was $13°C$ in London. What was the temperature difference between the two cities?

$-9°C$ to $13°C$ is a temperature difference of $22°C$

Practice – Unit 3 Directed number

1 Insert either $\boxed{>}$ or $\boxed{<}$ between the pairs of numbers to make the statements true:

(a) +4 $\boxed{}$ −7.   (f) −2 $\boxed{}$ +4.

(b) +7 $\boxed{}$ −3.   (g) −7 $\boxed{}$ −11.

(c) −11 $\boxed{}$ −4.   (h) +5 $\boxed{}$ +3.

(d) −5 $\boxed{}$ −13.   (i) −7 $\boxed{}$ 0.

(e) −5 $\boxed{}$ +2.   (j)   0 $\boxed{}$ 5.

2 The lowest temperature in Manchester one year was −9°C. The highest temperature that same year was 41°C above this. What was the highest temperature?

3 The bridge stands 10 m above the level of the surface of the river. The river bed is 5 m below the surface of the water. What is the distance between the bridge and the river bed?

4 In a game where arrows are fired at a target, you score (+4) points if you score a 'hit' and (−5) points if you score a 'miss'. What are the totals for these competitors?

(a) 4 'hits' and 1 'miss'

(b) 3 'hits' and 2 'misses'

(c) 2 'hits' and 3 'misses'

(d) 1 'hit' and 4 'misses'

5

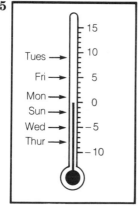

Fig. 7

Mr Green took the temperature daily at 8.00 a.m. His results are shown in Fig. 7.

(a) Work out the change in temperature daily, stating whether there had been a rise or fall and by how many degrees.

(b) If on Saturday there had been a fall of 9°C from the previous day then mark off Saturday's temperature.

6 Work out the value of these:

(a) $8-(-4)+(-3) = \dots$ (b) $-17-4 = \dots$

(c) $\dfrac{(-1)\times(-1)}{(+1)} = \dots$ (d) $\dfrac{-(-2)-(+3)}{(-2)} = \dots$

Aims of the unit

To revise:
1 The idea of a ratio,
2 Ratios in simplest form,
3 Direct and inverse proportion,
4 Model scales and map scales,
5 Bargains and best buys.

Ratios

A RATIO is the relation existing between two or more quantities of the same kind and there is very much a direct link between fractions and ratios.

$$2:6 = \tfrac{2}{6}, \; 5:15 = \tfrac{5}{15}, \; 8:24 = \tfrac{8}{24}$$

All these fractions cancel down to $\tfrac{1}{3}$. Hence the ratio in its simple form would be 1:3.

Since ratio is a comparison of sizes (i.e. lengths, masses, amounts of money etc.) no units are involved.

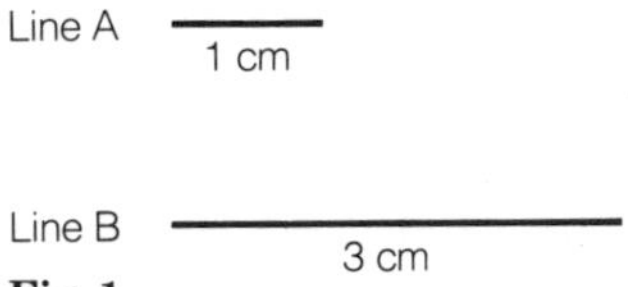

Line A ———
           1 cm

Line B ————————
             3 cm
Fig. 1

In Fig. 1 the ratio of Line A to Line B = 1 cm:3 cm
i.e. 1:3 (a comparison of the measurements –
units not required).

However, we must be certain that units are of the same type before we can compare them in ratio form.

Example What is the ratio of 1 m to 1 km?

The ratio would be 1 m:1 km, which is, of course, 1 m:1000 m, i.e. 1:1000.

Using a known ratio to solve a problem

K ▶ Provided that we are given enough information in the form of ratios and figures we can use the information to solve the problem.

Example 1 Divide £100 in the ratio 1:4.

Altogether five shares are required (i.e. 1 + 4).
Each share is worth $\dfrac{£100}{5} = £20$

Hence 1 share is £20 and four shares will be $4 \times £20 = £80$
Thus £100 divided in the ratio $1:4$ gives $£20:£80$

Example 2 In a class of 30 pupils the ratio of the number of boys to the number of girls is $7:8$. How many girls are there?

$$\text{Boys}:\text{Girls}$$
$$7:8$$

We need altogether 15 shares (i.e. $7+8$)

One share will be $\dfrac{30}{15} = 2$ pupils

Hence, since the girls account for eight shares, there must be $8 \times 2 = 16$ girls in the class (there would of course be 14 boys)

Example 3 The ratio of the length of a room to its width is $4:3$. How wide is a room that is 10 m long?

$$\text{Length}:\text{Width}$$
$$4:3$$

Hence four shares $= 10$ m, and so 1 share $= \dfrac{10}{4} = 2\frac{1}{2}$ m

Now width is three shares, i.e. $3 \times 2\frac{1}{2}$ m $= 7\frac{1}{2}$ m

NB: It is very important that ratios are always used in the correct order. For example, if $A:B = 2:5$ then $B:A = 5:2$ etc.

Direct proportion

Two quantities are in DIRECT PROPORTION if an increase (or decrease) in one quantity is matched by an increase (or decrease) in the same ratio in the second quantity. The type of problem can be seen in the following example.

Example 8 blocks of chocolate cost 96p. How much will
(a) 5 blocks of chocolate cost?
(b) 13 blocks of chocolate cost?

The problem is best solved by finding the cost of 1 chocolate block, i.e.

8 blocks cost 96p $\therefore$ 1 block costs $\dfrac{96}{8}$p $= 12$p

Hence, (a)   5 blocks cost 5×12p $= 60$p
and (b) 13 blocks cost 13×12p $= 156$p $= £1.56$

Inverse proportion

If an increase (or decrease) in one quantity produces a decrease (or increase) in a second quantity in the same ratio, then the two quantities are said to be in INVERSE PROPORTION.

For instance, if it takes 8 days for 2 men to do a job, how long would it take 4 men to do the same job?

When we study the problem closely, it is obvious that if more men do the same job then the job is completed in a much faster time – in this case 4 men will do the job in half the time, i.e. 4 days.

Remember to check your result to see if it is 'reasonable'. In the example above it is reasonable to assume that 4 men will do the job twice as quickly as 2 men, and on the other hand 1 man would take twice as long to complete the job, i.e. 16 days.

Scale

Models are made to SCALE. This means that a model is built in proportion to the original aircraft, ship, railway locomotive etc., and as a result looks very realistic.

Example A model car is produced $\frac{1}{50}$ th of the full size.

(a) What length on the model would be represented by 1 m on the real car?
(b) What length on the real car does 3 cm on the model represent?

> (a) Model : full size
>   1 : 50
>
> Hence, 50 cm on the real car is represented by 1 cm on the model
> and so, 100 cm (1 m) on the real car is represented by 2 cm on the model
>
> (b) 1 cm on the model represents 50 cm on the real car
> so,   3 cm on the model represents 50×3 cm on the real car
> $= 150$ cm ($1\frac{1}{2}$ m) on the real car

Map scales

A map is designed to represent on paper a particular section of land area. In order to represent the section of land accurately, the land measurements are scaled down to enable it to fit the paper it is printed on. The scale of the map is sometimes called the MAP RATIO. Maps are produced in a variety of scales. The smaller the map ratio, the smaller the area of land represented etc.

Example On the map the ratio is 1 : 200 000. What is
(a) the actual distance that is represented by a line 15 cm long?
(b) the distance on the map representing 10 km on the ground?

> (a) 1 cm on the map represents 200 000 cm on the ground so,
> 15 cm on the map represents $200\,000 \times 15$ cm on the ground
> $= 3\,000\,000$ cm $= 30\,000$ m $= 30$ km

(b) 10 km on the ground $= 10 \times 1000$ m
$$= 10 \times 1000 \times 100 \text{ cm}$$
$$= 1\,000\,000 \text{ cm on the ground}$$
now since the map scale is $1:200\,000$
$200\,000$ cm on the ground are represented by 1 cm on the map
$1\,000\,000$ cm on the ground are represented by 5 cm on the map

Bargains and best buys

We often have to decide when shopping which particular package of the same product is the better value for money. Usually the larger packet/container is better value; but this is not always the case. A quick calculation can soon reveal which is the BARGAIN.

Example A D.I.Y. store sells paint in three different sizes of can:
(A) 5 litres for £6.95, (B) 3 litres for £3.90, (C) 2 litres for £2.66.
(a) Which can represents the best value?
(b) What is the cheapest cost for 5 litres of paint?

(a) In order to find the can which offers the best value we need to find the cost per litre for each can.

$$A = \frac{£6.95}{5} = £1.39/\text{litre}$$

$$B = \frac{£3.90}{3} = £1.30/\text{litre}$$

$$C = \frac{£2.66}{2} = £1.33/\text{litre}$$

Hence the best value for money is Can B

(b) 5 litres – either 1 Can A *or* 1 Can B + 1 Can C
£6.95 *or* £3.90 + £2.65 = £6.55
Hence the cheapest cost for 5 litres is £6.55.

Practice – Unit 4 Ratio, proportion and scale

1 Express these ratios in their simplest form:
(a) $5:20$.　　(c) $12:15$.
(b) $4:12$.　　(d) $15:20:30$.
2 Simplify the following ratios
(a) 5 mm : 2 cm.　　(c) 700 m : 1 km.
(b) £6 : 50p.　　(d) 400 g : 2 kg.
3 Divide these amounts in the given ratio:
(a) £160 in ratio $3:2$.
(b) 120 cm in the ratio $2:3:5$.

4 Andy and Matt share a paper round. Andy does 3 days and Matt does 4 days each week. Andy earns £3.60. How much does Matt earn?

5 A sum of money is shared into three parts in the ratio 2:4:5. If the largest share is £55, what is the total amount of money shared?

6 A recipe for 15 cakes include 90 g of margarine and 300 g of flour.
(a) How much flour would be needed to make 20 cakes?
(b) How much margarine should be used to 120 g of flour?

7 Four men can do a particular job in 3 hours.
(a) Working at the same rate, how long will it take 1 man to do the same job?
(b) How long would it take 3 men to do the job?

8 The liner, submarine and train shown in Fig. 2 are to a scale of 1 cm:20 m. Calculate their actual lengths.

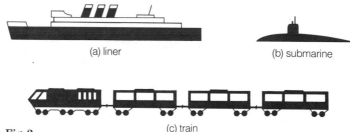

(a) liner   (b) submarine

Fig. 2   (c) train

*9 The scale on a map is 1:50 000.
(a) What distance is represented by a line on the map 8 cm long?
(b) What length of line on the map would represent an actual distance of 5 km?
(c) There is a lake that has an area of 2 km². What would the area be on the map that represents the lake?

*10 SHAMPOO   SHAMPOO   SHAMPOO

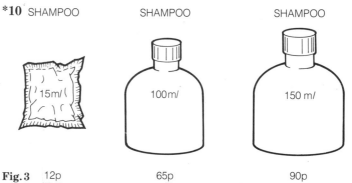

Fig. 3   12p   65p   90p

(a) Which size is best value?
(b) Describe how you reached your decision.

Unit 5 Approximation and estimation

Aims of the unit

To revise:
1 Approximating calculations,
2 Rounding-off,
3 Estimations,
4 Significant figures (sig. figs.),
5 Decimal places (dec. pl.).

Approximating calculations

For every calculation we perform, we should really carry out a rough check in order to satisfy ourselves that the result is reasonably accurate.

Example 1 How much money would I need to be able to buy 18 small presents which cost 19p each?

As an approximation we could easily find the cost of 20 articles costing 20p each, i.e.

$$20 \times 20 = 400p = £4.00$$

This answer is obviously too high (the actual answer is £3.42), but it does give us a rough idea of how much we should expect to have to pay.

Example 2 What would be the approximate mass of 32 screws each weighing 1·8 g?

We can roughly interpret this problem as being 30 screws weighing 2 g each, i.e.

$$30 \times 2 = 60 \text{ g (the actual answer is } 57·6 \text{ g)}$$

Rounding off

If a building contractor was asked to estimate the cost of alterations to a property, he would work out roughly what each aspect of the job would cost and add them together to supply the customer with an approximate cost.

The cost may have been rounded off to the nearest £100, £50 or the nearest £10.

Example 1 £358 rounded off to the nearest £100 would be £400.

Example 2 £235 rounded off to the nearest £50 would be £250.

Example 3 £83 rounded off to the nearest £10 would be £80.

Example 4 £25.68 rounded off to the nearest £1 would be £26.

NB: When an amount is exactly halfway between limits, it is accepted that we round up – to the limit above the amount.

Example 5 £2.50 rounded off to the nearest £1 would be £3.

Obviously we can apply the same theory to ordinary numbers and in the same way approximate these ordinary numbers to the nearest 10, 100, 1000 etc.

Estimation

It is very useful to be able to estimate the size of many things such as lengths of lines, angle sizes, heights of houses, trees etc.

Example 1

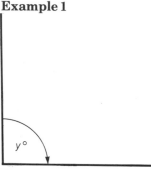

Fig. 1

An estimate of $y°$ would be 90°.

Example 2

A •————• B

Fig. 2

The approximate length of the line AB would be 1 cm.

Significant figures (sig. figs.)

Sometimes a number has far too many figures in it for practical use. This problem can be overcome by reducing the number to a certain number of significant figures. We can correct a number to 3 sig. figs., or 2 sig. figs., or 4 sig. figs., or however many significant figures we require.

The usual rule is that, if the digit to be discarded is greater than or equal to 5, then 1 is added to the previous digit; otherwise the previous digit is not changed.

To express a number to a given number of sig. figs. we count the total number of digits in the result.

Example 1 3·7609 is a decimal expressed to 5 sig. figs.

$\quad$ = 3·761 to 4 sig. figs. (the 9 changes 0 to 1)
$\quad$ = 3·76 to 3 sig. figs. (the 1 leaves the 6 unaltered)
$\quad$ = 3·8 to 2 sig. figs.
$\quad$ = 4 to 1 sig. fig. (or, in this case, to the nearest whole number)

Example 2 0·007 406 is a decimal expressed to 4 sig. figs. It is important to note that the '0's between the decimal point and the first non-zero digit (i.e. 7) are *not* significant.

Hence, 0·00741 (correct to 3 sig. figs.)
and   0·0074 (correct to 2 sig. figs.) etc.

In the case of whole numbers, suppose we consider 33 582 which might be the population of a small town.

Expressed to 4 sig. figs. it would be 33 580
Expressed to 3 sig. figs. it would be 33 600
Expressed to 2 sig. figs. it would be 34 000

Note that although the zeros are not significant they must not be left out – otherwise the whole sense of the approximation is lost. In this particular case we say that the approximate population of the town is 34 000.

Decimal places (dec. pl.)

In a similar way we can also approximate numbers by means of reducing the number of decimal places in the number. This is called CORRECTING the number to a specified number of decimal places.

Example Correct 3·0365 to (a) 3 dec. pl., (b) 2 dec. pl., (c) 1 dec. pl.

(a) 3·037 (to 3 dec. pl.)
(b) 3·04 (to 2 dec. pl.)
(c) 3·0 to 1 dec. pl.)

NB: In everyday life, we need to approximate to an answer that has a practical meaning. For instance, an answer to a problem of 31·473 cm would best be recorded as 31·5 cm (i.e. correct to 1 dec. pl.); and an answer of £33.827 would obviously be best recorded as £33.83 (i.e. correct to 2 dec. pl.), for it to have a sensible meaning.

Practice – Unit 5 Approximation and estimation

1 A TV is marked £159.95, a watch £23.95 and a calculator £12.95.
(a) What is the approximate cost of each item, to the nearest £10?
(b) What is the approximate cost of each item, to the nearest £1?
2 Approximate these measurements to 1 decimal place.
(a) 25·34 secs.
(b) 10·09 secs.
(c) 13·66 metres.

3 Estimate the following:

(a) In Fig. 3

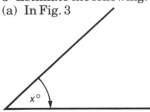

Fig. 3

angle x = ...°.

(b) In Fig. 4

A ——————————— B

Fig. 4

length AB = ... cm.

(c) The length of a family car in metres.

(d) The width of a railway carriage.

4 Examine the number 54 975, which is the exact attendance at a soccer international. Write an approximation of the attendance:

(a) Correct to the nearest ten thousand.

(b) Correct to the nearest thousand.

(c) Correct to the nearest hundred.

(d) Correct to the nearest ten.

*5 Examine the number 2408·745. Write an approximation of this number:

(a) Correct to 2 decimal places.   (d) Correct to 2 sig. figs.

(b) Correct to 1 decimal place.   (e) Correct to 4 sig. figs.

(c) Correct to 3 sig. figs.

*6 Choose the correct measurement from the set {20 cm, 15 cm, 10 cm} that corresponds to the missing side lengths in the triangles of Fig. 5.

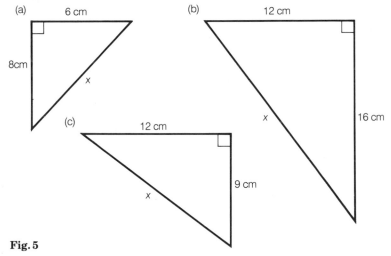

Fig. 5

Aims of the unit

To revise the basic operations of the calculator, i.e.

$$+, \ -, \ \times, \ \div, \ \sqrt{}$$

Calculators

Modern school mathematics, leading to all GCSE examinations, is becoming more and more dependent on the use of the electronic calculator. Much of the work covered will depend on your basic skill with the calculator.

Unfortunately, the manufacturers of calculators have done very little to standardize the layout of the keyboards on their models. So, although all the calculators perform more or less the same functions, the positions of the keys may vary quite considerably.

To overcome this problem it is a good idea to familiarize yourself (with the aid of the instruction manual) with your own calculator and discover not only what it will do, but also how it does it. Any spare time you may have is well spent experimenting with your machine to discover its capabilities and limitations.

NB: If your course includes trigonometry then you will need a scientific calculator.

To be sure that you can handle your calculator for the basic skills required, work through the following selection of questions.

They are designed to help you improve your calculator skills.

Fig. 1 Diagram of a calculator

Practice – Unit 6   Calculators

1 Complete Table 1 below with the help of your calculator:

Table 1

Sum	Estimate	Calculator
28×11 42×97 103×78 61×121	$30 \times 10 = 300$	308

2 Follow these instructions:

   (i) Enter a three-figure number in your calculator.

   (ii) Make it a six-figure number by repeating the three digits used in (i), e.g. 625 625.

   (iii) Divide by 7.

   (iv) Divide by 11.

   (v) Divide by 13.

What do you notice about your answer?

3 Multiply the numbers in the circles to give the answer inside the square between them, for example as in Fig. 2.

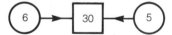

Fig. 2

Complete these:

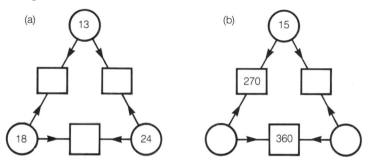

(c)

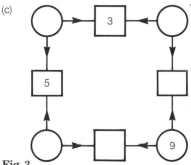

Fig. 3

4 A supermarket stacks tins as in Fig. 4.
(a) How many tins will be in a stack of 10 layers?
(b) How many tins will be in a stack of 15 layers?
(c) How many layers will be needed to stack 210 tins?
(d) How many tins will be in the bottom layer of the stack in (c)?

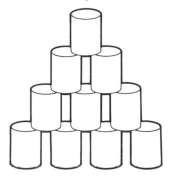

Fig. 4

5 Complete the square roots of Table 2. Answer to 3 decimal places.

Table 2

Sum	Guess	Calculator
$\sqrt{10}$		
$\sqrt{20}$		
$\sqrt{50}$		
$\sqrt{100}$		

Unit 7 Units – metric and Imperial

Aims of the unit

To revise all units of mass, length, area, volume and capacity in common use.

Mass

The metric system of mass used in most countries is as shown in Table 1 with abbreviations mg for milligrams, g for grams and kg for kilograms.

Table 1

1000 mg = 1 g
1000 g = 1 kg
1000 kg = 1 tonne (metric)

Since we are using some Imperial measurements in this country, · Table 2 shows a few in common use.

Table 2

16 ounces = 1 pound
14 pounds = 1 stone
112 pounds = 1 hundredweight
20 hundredweights = 1 ton (Imperial)

Since we know how to handle all decimal operations, metric mass problems should be no problem for us.

However, it is of course very important to be sure that the units in which we are working are the same – otherwise they must be converted before the problem is attempted.

Length

Most countries of the world now use a metric system of measuring for length. It is based on the following units.

Using the abbreviations mm for millimetre, cm for centimetre, m for metre and km for kilometre, we have Table 3.

Table 3

10 mm = 1 cm
100 cm = 1 m, also 1000 mm = 1 m
1000 m = 1 km

Since we do still use some Imperial units in this country, these are listed in Table 4.

Table 4

12 inches = 1 foot	
3 feet = 1 yard	220 yards = 1 furlong
	(used in horse racing)
and 1760 yards = 1 mile	8 furlongs = 1 mile

Once again, we should easily be able to handle operations involving metric units of length. But ensure that the units are the same before starting any operation.

Capacity

Capacity is a measure of the amount of SPACE that a liquid or other fluid takes up within the container in which it is held.

Metric units of capacity are shown in Table 5.

Table 5

10 millilitres (ml)	= 1 centilitre (cl)
100 cl	= 1 litre (l)
1000 ml	= 1 litre (l)

Once again, in this country, we are still using some of the Imperial units of capacity, as shown in Table 6.

Table 6

20 fluid ounces (fl. oz.)	= 1 pint (pt)
2 pints	= 1 quart (qt)
4 quarts	= 1 gallon (gall)
or 8 pints	= 1 gallon

As for length and mass, the metric operations are very simple, providing we start with all units of the same type.

Area

Common metric units of area are shown in Table 7. Notice that units of area are 'square' units.

Table 7

$100 \text{ mm}^2 = 1 \text{ cm}^2$
$10\,000 \text{ cm}^2 = 1 \text{ m}^2$
$1\,000\,000 \text{ m}^2 = 1 \text{ km}^2$

Volume

Common metric units of volume are shown in Table 8. Notice that units of volume are 'cubic' units.

Table 8

$$1000 \text{ mm}^3 = 1 \text{ cm}^3$$
$$1\,000\,000 \text{ cm}^3 = 1 \text{ m}^3$$

Once again, we must be very careful to ensure that, before calculations are performed, all units are the same. In any conversions we must be very careful with the accuracy etc.

Practice – Unit 7 Metric units

1 Complete these changes of units:
(a) 5 m   = ... cm   (b) 3 km   = ... m
(c) 8 cm   = ... mm   (d) 4·25 kg = ... g
(e) 2·75 l   = ... ml
2 Mrs Jones's rubber plant is 55 cm tall. Mrs Smith's plant is 1·7 m tall.
(a) How many cm taller is Mrs Smith's plant?
(b) About how many times taller is it?
3 An 8 m length of wood is cut to make posts 150 cm long.
(a) How many complete posts can be cut?
(b) What length of wood will be left over?
4 The diameter of a penny is 2 cm. (a) How many pennies placed side by side will make a distance of 5 m? (b) How far will £4 worth of pennies reach?
5 Complete these changes of units:
(a) 650 ml   = ... l.   (b) 2400 g = ... kg.
(c)   42 mm = ... cm.   (d) 25 cm   = ... m.
(e) 600 m   = ... km.
6 A housewife returns from the supermarket with:
  3 bags of carrots, each of mass 2·2 kg.
  8 cans of beans, each of mass 425 g.
  4 packets of cereal, each of mass 440 g.
  3 jars of jam, each of mass 340 g.
Calculate the total mass of her purchases in kg.
***7** Twenty-four tins of meat are packed into a carton and the total mass is 16 kg. If the carton alone weighs 400 g calculate the mass of 1 tin of meat.
***8**

5 miles is approximately 8 kilometres

In Fig. 1 how far is it in km to:
(a) Dundee,
(b) Brechin,
(c) Perth?

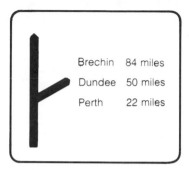

Fig. 1

Unit 8 Money

Aims of the unit

To revise:
1 Foreign currency and holidays,
2 Hire purchase, discount and VAT,
3 Simple and compound interest,
4 Household bills,
5 Wages,
6 Profit and loss.

Foreign currency and holidays

When we travel abroad on holiday we have to change our money (pounds sterling) into the currency of the country we wish to visit. The pound varies from day to day against other currencies. But we can find a list of EXCHANGE RATES published daily in newspapers and displayed at banks.

Table 1 shows a possible day's value of the pound against a selection of other currencies. From these figures it is easy to work out how much we could expect to exchange our money for.

Table 1

France	9·40	francs = £1	(sterling)
Germany	2·78	marks = £1	(sterling)
Spain	178	pesetas = £1	(sterling)
Switzerland	2.50	francs = £1	(sterling)
USA	1·60	dollars = £1	(sterling)

Example 1 Jane is going on a trip to France and wishes to change £15 into francs. How many francs will she receive?

£1 is equivalent to 9·40 francs

∴ £15 is equivalent to 9·40 × 15 francs
= 141 francs

In a similar way people visiting this country wish to change their currency into pounds (sterling) and knowing the exchange rate involved it is quite easy to find how much they would expect to get.

Example 2 Hans is travelling from Germany to London for his holidays, he wishes to exchange 200 DM for pounds sterling, how much does he receive?

2·78 DM is equivalent to £1 (sterling)

1 DM is equivalent to $\dfrac{£1}{2·78}$

Hence 200 DM is equivalent to $\dfrac{£1}{2·78} \times 200 = £71.94$

Changing from one foreign currency to another

In order to do this we need to know the exchange rates of the two currencies with respect to pounds sterling.

Example Change 500 pesetas into Swiss francs.

178 pesetas = £1

∴ 1 peseta = $\dfrac{£1}{178}$

Hence 500 pesetas = $\dfrac{£1}{178} \times 500 = £2.81$

We then change £2.81 into Swiss francs

£1 = 2.50 Swiss francs

∴ £2.81 = 2.50 × 2·81 = 7.03 Swiss francs

Hence 500 pesetas is equivalent to 7.03 Swiss francs

Conversion graphs

Another way of converting currencies is by means of a conversion graph, as shown in Fig. 1 which compares pounds sterling with Swiss francs.
NB: Results will only be approximate; the accuracy depends on the size of the scale of the graph.

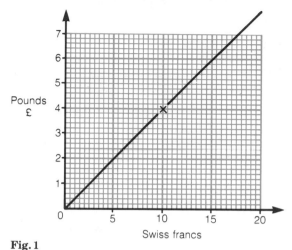

Fig. 1

Example 1 What is the value of 16 Swiss francs in pounds sterling? (Approximately £6.40.)

Example 2 What is the value of £3 in Swiss francs? (Approximately 7.5 Swiss francs.)

Practice – Unit 8 Foreign currency and holidays

1 The exchange rates in these countries were:
   USA: £1 = 1.6 dollars   Spain: £1 = 178 pesetas
   France: £1 = 9.40 francs   Germany: £1 = 2.78 marks
(a) What is £5 worth in Spain?
(b) What is £10 worth in the USA?
(c) What is 147 marks worth in pounds sterling?
(d) What is 42 dollars worth in pounds sterling?
(e) How many Spanish pesetas is 333 francs worth?

2 A pair of shoes is marked at 35 dollars. How much is that in pounds sterling?

3 A handbag is priced at 816 pesetas. How much is that in pounds sterling?

4 Perfume is on sale in the duty-free shop at 133.2 francs. On the plane the same perfume is sold at £27.50. Which is the better buy and by how much?

5 In the same holiday resort hotels have the different tariffs shown in Table 2.

Table 2

Hotel	10-day holiday
Don Juan	£125
Luxe	£150
Sol	£180

(a) What is the cost per day at the Hotel Luxe?

(b) What is the difference in price between the dearest and cheapest holiday for a 10-day stay for two adults?

(c) If children get a 25% reduction, what is the cost of a 10-day holiday at the Hotel Sol for a married couple and their three children?

(d) Mr Hill goes to the hotel Don Juan where he is charged at an exchange rate of £1 = 178 pesetas. How much does he pay in pesetas for a 10-day holiday?

*6 The exchange rate in Paris is 9.40 francs to the pound and in Frankfurt it is 2.78 marks to the pound.

(a) A man changes £200 into francs in Paris. How many francs does he get?

(b) After spending all his francs, he changes £150 into marks in Frankfurt. How many marks does he get? Answer to the nearest mark.

(c) He spends 300 marks and then changes the rest back into pounds sterling. How many pounds does he get? Answer to the nearest pound.

*7 Fig. 2 is a graph to compare dollars ($) and pounds (£).

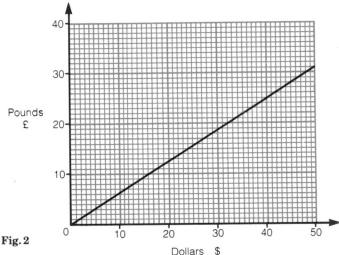

Fig. 2

Use the graph to answer:
(a) How many £s is 30 dollars worth?
(b) How many dollars is £13 worth?
(c) A camera costs £25 in London and $32 in New York. What is the difference in price in £s?

Hire purchase

Hire purchase allows a person to obtain goods without paying the total price immediately. The payments are usually spread over a period of time up to 3 years and may be paid weekly, monthly etc. More often than not, a DEPOSIT is paid at the time of purchase and then interest is charged on the BALANCE remaining to be paid.

Hence, the HP price is almost certainly to be higher than the cash price.

Example 1 A hi-fi system has a cash price of £280 but on HP terms a deposit of 20% is required and 24 monthly payments of £11.85 must be made. How much more than the cash price is the HP price?

$$\text{For HP 20\% of £280} = \text{deposit} = \frac{20}{100} \times £280 = £56 \text{ deposit}$$

$$24 \text{ payments of } £11.85 = £284.40$$

$$\text{Hence total HP cost} = £284.40 + £56 = £340.40$$

HP price is £60.40 more than the cash price

Discount

Discount is an amount knocked off the cash price as a means of attracting customers. It is often expressed as a percentage of the usual selling price and, for example, is displayed as '20% off while stocks last'.

Example A pair of shoes priced at £23.50 is offered in a sale at a discount price of 20% off. How much will the shoes cost?

$$20\% \text{ of } £23.50 = \frac{20}{100} \times 23.50 = £4.70$$

Hence discount is £4.70 and so sale price $= £23.50 - £4.70$
$$= £18.80$$

Also, if we regard the original price as 100% and the discount is 20% then it is clear that the selling price will be 80% of the old price.

$$\text{To find 80\% of £23.50 we find } \frac{80}{100} \times £23.50 = £18.80$$

It is possible to find the cash price if we are given the sale price and the discount.

Example A record is offered in a sale at £4.95 after a 10% discount. What is the normal cash price?

10% discount, hence £4.95 is equivalent to 90%

and so $\dfrac{£4.95}{90}$ is equivalent to 1%

$\therefore$ cash price (100%) is equivalent to $\dfrac{£4.95}{90} \times 100 = £5.50$

Obviously this calculation is very easily performed using a calculator.

VAT (value added tax)

K

Value added tax is a government tax paid on certain goods such as furniture, electrical goods, cars etc. The rate of taxation in 1990 is 15% but it could be changed at any time by the Chancellor of the Exchequer in his annual budget.

Example 1 Mr Gibson buys a car costing £5100 + VAT. If VAT is due at the rate of 15%, how much will Mr Gibson have to pay?

$$15\% \text{ of } £5100 = \dfrac{15}{100} \times £5100 = £765$$

Hence total cost of the car = £5100 + £765 = £5865

Example 2 A washing machine costs £345 including VAT. With VAT at the rate of 15%, what is the pre-VAT price of the machine?

115% is equivalent to £345 total price

1% is equivalent to $\dfrac{£345}{115}$

So the pre-VAT price (100%) is equivalent to

$\dfrac{£345}{115} \times 100 = £300$

Example 3 The VAT paid on an article amounts to £90 (VAT = 15%). What is the total price of the article (including VAT)?

£90 is equivalent to 15%

$\therefore \dfrac{£90}{15}$ is equivalent to 1%

Hence the pre-VAT cost (100%) is equivalent to

$\dfrac{£90}{15} \times 100 = £600$

and so total price paid = £600 + £90 = £690

Practice – Unit 8 Hire purchase, discount and VAT

1 Complete Table 3.

Table 3

Item	Deposit	Instalments	HP Price
Bed	£15	12 months @ £13.50	
TV	£20	24 months @ £12.50	
Hi-fi	£25	30 months @ £6.75	

2 A radio was £40 but is marked 'Now! 15% Off' in a sale. A £36 football is marked '20% off'.
What are the sale prices?

      Radio = £... .   Ball = £... .

3 A restaurant advertises a Sunday lunch at £4.50 per head plus 15% VAT.
(a) What is the total cost for Sunday lunch for 4 people?
(b) What is the cost per person:

4 Complete the missing items in the following account:

    ACORN GARAGE
    Labour 8 hours at £8 = £
    Parts used (exhaust) + £24.00
                      _____
              Total =
       15% VAT =
                      _____
       Total bill =
                      _____

*5 The VAT payable on a record is 75p.
(a) What does the record cost without VAT?
(b) What does the record sell for?

*6 Jim bought a second-hand car and paid a deposit of $33\frac{1}{3}\%$ of the cash price. He then paid 30 monthly payments of £95.50. The deposit amounted to £1250. Find:
(a) The cash price.
(b) The total HP price.

*7 A video can be bought for £660 cash or on HP with a deposit of 30% of the cash price and 12 equal monthly payments. Interest is charged at 15% on the balance.
(a) What is the deposit?
(b) What is the balance?
(c) What is the balance outstanding *plus* interest charges?
(d) What do the monthly payments work out at per month?

Simple interest

Simple interest is the interest paid on a personal loan from a bank. The amount of simple interest paid depends on the amount borrowed, the time taken for repayment and the current rate of interest per year (per annum or p.a.)

Example If we borrow £1000 for 2 years at a rate of 10% p.a.

> Then £1000 (PRINCIPAL or amount borrowed)
> +10% of £1000 interest for 1st year = £100
> +10% of £1000 interest for 2nd year = £100
> ∴ total repayment = £1200
> Hence simple interest is £200

Simple interest can be calculated using the formula

$$I = \frac{PRT}{100} \text{ where } \begin{aligned} I &= \text{simple interest} \\ P &= \text{principal (amount borrowed)} \\ R &= \text{rate \% p.a.} \\ T &= \text{time (years)} \end{aligned}$$

Example 1 Calculate the simple interest on a loan of £250 at 15% for 6 years.

$$P = £250, R = 15\% \text{ p.a.}, T = 6 \text{ years}$$
$$I = \frac{PRT}{100} \therefore I = \frac{£250 \times 15 \times 6}{100} = £225$$

So the simple interest would be £225

It is possible to find also the principal, the rate or the time provided we are given the other three items of information:

$$P = \frac{100I}{RT}, \quad R = \frac{100I}{PT}, \quad T = \frac{100I}{PR}$$

Example 2 A man invested £400 in a bank and after 5 years his investment had increased to £560. What was the rate of simple interest?

$$P = £400, \ I = £160 \ (£560 - £400), \ T = 5 \text{ years}$$
$$R = \frac{100I}{PT}, \ \therefore R = \frac{100 \times 160}{400 \times 5} = 8\%$$

∴ Rate of simple interest = 8% p.a.

Compound interest

When compound interest is calculated, each year's interest is added to the amount of the loan/investment before next year's interest is calculated. Hence, the interest is actually increasing each year (it includes interest on the interest, as well as on the principal).

Example 1 Find the compound interest on £100 borrowed for 3 years at 10% p.a.

In order to find this amount we have to calculate the interest for the first year and add this to the principal. Then we have to repeat the operation on this new principal, and so on.

> For the first year: the interest will be 10% of £100 = £10
> For the second year: principal = £100 + £10 = £110
> interest will be 10% of £110 = £11
> For the third year: principal = £110 + £11 = £121
> interest will be 10% of £121   = £12.10

∴ In 3 years the compound interest will be
£10 + £11 + £12.10 = £33.10

Example 2 A man invests £500 for 3 years at 8% p.a. compound interest. Find:

(a) The total amount of the investment after 3 years.
(b) The total interest earned during the 3 years.

Interest in 1st year $= 8\%$ of £500 $= \dfrac{8}{100} \times £500 = £40$

Principal for 2nd year = £500 + £40 = £540

Interest in 2nd year $= 8\%$ of £540 $= \dfrac{8}{100} \times £540 = £43.20$

Principal for 3rd year = £540 + £43.20 = £583.20

Interest in 3rd year $= 8\%$ of £583.20 $= \dfrac{8}{100} \times £583.20$
$= £46.66$ (to the nearest penny)

(a) Hence, total investment after 3 years = £583.20 + £46.66
$= £629.86$
(b) and the total interest earned = £629.86 − £500 = £129.86

Practice – Unit 8 Simple and compound interest

1 A lady invests £800 in a savings scheme that pays simple interest at the annual rate of 12%.
(a) What is the interest due after 1 year?
(b) What will her interest amount to if she leaves her money in for 5 years?

2 Twins Matthew and Victoria save in different banks. Matthew has £96 invested where the rate of interest is 10% per annum, simple interest. Victoria has £90 invested where the simple interest rate is 9% per annum.

(a) What does Matthew have in the bank, in total, after 1 year?

(b) What does Victoria have in total after 1 year?

(c) Who has the most after 5 years?

*3 A retired man wishes to have an income of £1000 per annum. How much must he invest in a building society paying $12\frac{1}{2}\%$ interest per annum?

*4 A boy receives a legacy of £80 and puts it into a bank paying simple interest at a rate of 11% per annum. After how many years will he have received £44 interest?

*5 A person borrows £2400 from a moneylender who charges compound interest of 10%. How much does he pay altogether if he takes 3 years to repay the loan?

*6 A man has £400 to invest for 3 years. He can invest it at a simple interest rate of $12\frac{1}{2}\%$ p.a. or compound interest of 10% p.a.

(a) Which is the better investment?

(b) By how much is it better?

Household bills

Most householders receive bills for services such as electricity, gas, telephone and rates.

In the case of gas and electricity we have meters in our houses measuring the amount of gas and electricity we use. These meters are periodically read and a bill produced by the respective companies.

Meters are generally of two types: (a) dials (Fig. 3), (b) digitals (Fig. 4).

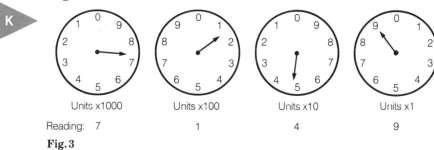

Units ×1000	Units ×100	Units ×10	Units ×1
Reading: 7	1	4	9

Fig. 3

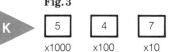

Reading: 5 4 7 8

Fig. 4

Whichever type of meter we have, the number of units used is calculated by finding the difference between the previous reading and the present reading.

Example Previous reading 1469; present reading 2103.

units used = 2103 − 1469 = 634 units

This information is displayed on the bill, together with the cost per unit and hence the cost of the gas/electricity used. To this is added a STANDING CHARGE (a set amount charged to the customer for administration costs). The final total to be paid is set out at the bottom of the bill.

Telephone bills

Telephone bills are very similar to the above, except that the meters are situated at the local exchange and, instead of a standing charge, the customer is charged a set amount known as RENTAL.

Any calls which are made through the operator are noted on the bill and the cost of these units is then added to the cost of calls made directly to give the total bill. VAT at 15% is then added to give the amount to be paid.

Community charge (poll tax)

Another bill we pay is the community charge. This is a tax on individuals aged 18 years and above. It is collected by local authorities for services such as education, social services, police, fire, libraries, highways and refuse collection. The amount of the community charge is fixed annually by each local authority in order to meet its proposed spending. Some people i.e. pensioners, the unemployed, single parents, the disabled and low earners, may qualify for a possible 80% reduction in their poll tax.

Example Calculate the community charge for a pensioner granted an 80% reduction on the standard charge if this authority has set this charge at £313 p.a.

He will therefore pay 20% of £313

$$\text{Amount to pay} = \frac{20}{100} \times £313 \times £62.60$$

Uniform business rate

All business premises have a rateable value. For 1990/91 a business rate has been fixed at 34.8 pence per pound for England and Wales. This means that for every pound of rateable value on business premises the owner has to pay 34.8p into the national pool.

Example How much would business premises with a rateable value fixed at £7000 have to pay in business rates?

Amount to pay = 7000 × 34.8p = £2436

Practice – Unit 8   Household bills

1 Read the meters shown in Fig. 5.

(a)

(b)

(c)

Fig. 5   ×1000   ×100   ×10

2 Work out how many units have been used if the meter readings are as shown in Fig. 6.

Previous reading   Present reading

(a)   5 6 7 8   5 7 8 9

(b)   4 0 9 0 6   4 1 3 7 8

(c)   4 1 3 8 4   4 3 2 7 9

Fig. 6

3 Complete the electricity bill given in Table 4.

Table 4

Meter readings			Amount
Previous	Present		
57776	59595	Units used = ... at 6p per unit Standing charge	= £.... = £6.80
		Total now due	= £...

4 Electricity is charged at 6.5p per unit and the standing charge is £6.80. Mrs Williams has her meter read and the reading is 2036 units. If the previous reading was 1820, work out her electricity bill for the quarter.

5 Complete the gas bill given in Table 5.

Table 5

Meter readings		Gas supplied	Charges
Present	Previous	Units	£
5339	5271	... at 38p each Standing charge	... 10.60
		Total now due	£...

*6 Mr Barlow elects to pay his gas bill by a monthly bank standing order of £34 per month. After one year he has used the following:

Winter quarter	2254 therms
Spring quarter	1345 therms
Summer quarter	834 therms
Autumn quarter	1296 therms

Throughout the year, gas is charged at 8p per therm and quarterly standing charges are £3.80.

(a) How many therms does he use in the year?
(b) What is his total gas bill for the whole year?
(c) Was his monthly standing order sufficient?
(d) By how much was the estimated cost out?

*7 On 15 October the dials on my meter stood as shown in Fig. 7.

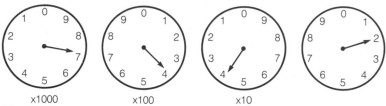

x1000   x100   x10

Fig. 7

(a) State the reading in figures.

(b) Between 15 October and 15 January the following year I used 420 units of gas. State the meter reading on the 15 January in figures.

(c) The number of therms of gas used is obtained by using the formula:

$$\text{Therms used} = \frac{(\text{Number of units used}) \times 105}{100}$$

420 units of gas is how many therms?

(d) Standing charge is £10.60 and gas is 38p per therm. Complete the gas bill.

*8 Table 6 shows telephone charges for 3 months.

Table 6

Rental	£18.50
Cost of metered units	£55.35
Cost of trunk call via operator	£ 1.15

(a) Metered units cost 5p. How many units were used?

(b) Calculate the total cost of the rental, metered units and trunk call.

(c) If VAT is charged at 15%, calculate the total amount of the bill.

*9 Gas charges may be worked out under two different tariffs as shown in Table 7.

Table 7

	General tariff	Gold Star
Standing charge	£4.25	£8.50
Charge for gas	First 100 therms 31.5p per therm Further therms 25p per therm	22p per therm

(a) Calculate the gas bill under the general tariff if 160 therms are used in a quarter.

(b) Calculate the gas bill under the Gold Star tariff if 90 therms are used.

(c) Calculate the smallest whole number of therms which will have to be used per quarter if the total cost under the Gold Star tariff is less than the total for the general tariff.

*10 A local city council sets the community charge for its area at £340.50.

(a) How much will a person with a 60% reduction pay?

(b) How much will a person with an 80% reduction pay?

The uniform business rate for 1990/91 is fixed at 34.8 pence per pound.

(c) How much will business premises with a rateable value fixed at £17 500 have to pay?

Wages – hourly paid work

Many people are paid for each hour they work and the wage is calculated by multiplying the hourly rate by the number of hours worked.

Example A labourer works a 39-hour week and is paid £2.80 per hour. What is his basic weekly wage?

Weekly wage = 39 × £2.80 = £109.20

But if this same labourer works extra hours above his basic 39 hours then he is usually paid OVERTIME which is paid at time-and-a-half ($1\frac{1}{2}$ times his normal hourly rate) or double-time (twice his normal hourly rate).

Thus, if the same labourer worked 46 hours (with overtime at time-and-a-half) he would be paid his basic weekly wage of £109.20 (as shown above) plus 7 hours overtime at $1\frac{1}{2}$ × £2.80 per hour, i.e.

7 × (£2.80 + £1.40) = 7 × £4.20 = £29.40

Hence his total wage for that week would be

£109.20 + £29.40 = £138.60

Wages – piece work

Piece work rates are paid according to the amount of work produced, not the time taken to do it. Many people working at home are paid piece work rates.

Example Elizabeth is paid £3.50 for every 100 pottery flowers she makes. How much does she receive for producing 1200?

She will receive 12 × £3.50 = £42

Sometimes people receive a bonus for producing amounts above their normal quota. For example, if Elizabeth was paid a bonus of £2 for every 50 she produced over 1200, how much would she receive in total for producing 1650?

> For 1200 she would receive £42 (as shown above)
> For an extra 450 she would receive $9 \times £2 = £18$
> Hence her total pay would be $£42 + £18 = £60$

Salary

Some jobs are paid on an annual basis, e.g. £9600 p.a. (each year). People who are paid a salary are given their money in 12 equal parts, paid monthly. Hence £9600 p.a. would be paid as £800 per month.

Example A bank clerk's salary is fixed at £7800 p.a. What would her monthly pay cheque be?

$$\text{Monthly pay cheque} = \frac{£7800}{12} = £650/\text{month}$$

Commission

Payment by commission is usually in the form of a fairly low basic wage plus a percentage of sales made by the person. The idea behind this is to encourage the salesperson to sell.

Example A car salesman is paid a basic wage of £2 per hour for a 40-hour week plus a commission of 3% on all sales. What would his total wage for the week be if he sold cars to the value of £4200 in that week?

> Basic wage $= 40 \times £2 = £80$
>
> Plus commission of 3% of $£4200 = \frac{3}{100} \times £4200 = £126$
>
> Thus total wage for week $= £206$

Deductions/stoppages

A job may be advertised at £130 per week. This is known as the GROSS PAY (pay before any deductions are made). The actual take-home pay, known as NETT PAY, will be the gross pay less deductions of income tax.

Income tax

This is the major stoppage and is collected by the worker's employer. Tax rates are set each year by the Chancellor of the Exchequer who decides on the personal allowances that are going to be given to tax payers. The personal allowance is the amount of money that a single or married person can earn before paying any tax. For 1990/91 the allowances were

single person or married woman: £3005; married man: £4725.

Any money earned above that amount is known as 'taxable income' and 25% (the tax rate for the same period) of this amount is taken away in tax.

Example A married man earns £9500 per year. Find (a) his taxable income, (b) income tax paid, (c) his monthly nett pay.

(a) Taxable income = Gross pay − Personal allowance
$$= £9500 − £4725 = £4775$$

(b) Income tax paid = 25% of £4775

$$= \frac{25}{100} \times £4775 = £1193.75$$

(c) His nett annual pay = Gross pay − Tax paid
$$= £8306.25$$

$$\therefore \text{ His nett monthly pay} = \frac{£8306.25}{12} = £692.19 \text{ per month}$$

Practice – Unit 8 Wages

1 A building worker is paid £3.50 an hour for a 40-hour week. What is his basic weekly wage?

2 A plumber whose basic working week is 35 hours earns £168. What is his hourly rate?

3 A painter gets paid £3.40 per hour for a basic 40-hour week. Overtime is paid at time-and-a-half.
(a) What would he get for an hour's overtime?
(b) What is his basic weekly wage?
(c) What would his wages be if he did 45 hours in one week?

4 A machine operator makes metal discs and is paid 25p for every 10 discs made. His daily output one week is:
Mon: 1000; Tues: 1400; Wed: 1520; Thur: 1850; Fri: 980.
(a) How much does he earn each day?
(b) How much does he earn in the week?

5 Mr Jones's salary was £12 600 per annum. When he retired he was paid a pension of 9/20 of this salary.
(a) What was his monthly pay cheque when he was working?
(b) What was his annual pension?

6 A shop assistant is paid a basic wage of £2.20 per hour for 35 hours *plus* a commission of 2% on all her sales. What would her total weekly wage be if she sold £650 worth of goods?

*7 The basic rate of pay is £3.50 per hour. The basic week is 40 hours and overtime is paid at time and a half.

(a) Calculate the basic weekly wage.

(b) Each week Sally works 45 hours. Calculate her weekly wage.

(c) In one year Mrs Harris earned £7525 without overtime. How many hours does this represent?

(d) In the same year Mrs Wilson earned £8288 but had 4 weeks holiday at the basic weekly wage and worked at least 40 hours per week in the remaining 48 weeks. What was the average number of hours she worked per week for these 48 weeks?

*8 A salesman received a basic salary of £6500 plus a commission of 20% of the profits made by the shop. His total salary was £10 750.

(a) Calculate his share of the profits.

(b) Calculate the total profits made by the shop.

His taxable income was £7800.

(c) Calculate his non-taxable allowances.

(d) If tax was paid at the rate of 25% on his taxable income how much tax did he pay?

(e) How much tax did he pay per month?

Profit and loss

If a shopkeeper buys something and then sells it for more than she bought it for – then she makes a profit. However, if she sells it for less than she bought it for – then she makes a loss.

$$\text{Hence Selling price} > \text{Cost price} \Rightarrow \text{Profit}$$
$$\text{but Selling price} < \text{Cost price} \Rightarrow \text{Loss}$$

A profit/loss is often written as a percentage of the cost price.

K

$$\% \text{ Profit/Loss} = \frac{\text{Actual profit}}{\text{Cost price}} \times 100$$

Example 1 A shopkeeper buys some goods for £50 and sells for £65.

(a) What is her profit?

(b) What is her percentage profit?

(a) Actual profit = £65 − £50 = £15

(b) $\% \text{ Profit} = \dfrac{\text{Actual profit}}{\text{Cost price}} \times 100 = \dfrac{15}{50} \times 100 = 30\% \text{ Profit}$

If the shopkeeper wishes to make a certain percentage profit on her sales then she must set her selling prices to give her this percentage profit.

Example 2 A trader buys some pottery for £105 and wishes to make 40% profit. What must he sell the pottery for?

Let the cost price be equivalent to 100%
i.e. £105 is equivalent to 100%
He requires a selling price equivalent to 140% and so,
since $100\% \equiv £105$

$$1\% \equiv \frac{£105}{100}$$

$\therefore 140\%$ selling price $\equiv \dfrac{£105}{100} \times 140 = £147$

$\therefore$ selling price must be £147

Similarly if the salesperson fixes a selling price for an article based on a certain percentage profit then the buying price (cost price) can be calculated as shown in the following example.

Example 3 An electrical contractor sells a TV for £240 making a 20% profit. What was the cost price of the TV?

Selling price $= £240 \equiv 120\%$

$\therefore 1\% \equiv \dfrac{£240}{120}$

$\therefore$ cost price $(100\%) \equiv \dfrac{£240}{120} \times 100 = £200$

NB: It is very important when using this method always to let the cost price be equivalent to 100%.

Practice – Unit 8 Profit and loss

1 An article is bought for £5 and sold for £3.
(a) What is the loss on the sale?
(b) What is the loss as a percentage of the cost price?
2 A greengrocer buys a box containing 200 oranges for £20 and sells them for 16p each.
(a) What is the profit per orange?
(b) What is the total profit if he sells all of them?
(c) What is his percentage profit?
3 A dealer buys an article for £40 and wishes to make a profit of 15%. How much should she sell it for?
4 A bicycle is bought for £80 and sold second-hand 12 months later. The owner does not want to make more than a 30% loss on the sale. What is the least he can sell the cycle for?
*5 A bicycle is sold for £133 and the dealer has made a profit of 40%. What did the dealer pay for the bicycle?

*6 A woman buys 150 articles at £8.50 each and sells them for £10.20 each.

(a) Calculate her total profit.

(b) Calculate this profit as a percentage of the buying price.

(c) The buying price increases by 8% but the selling price remains unaltered. How many articles does she now have to sell to make the same profit?

*7 A shopkeeper buys a coat for £36 and puts a price tag on it to give him a 30% profit.

(a) Calculate the marked price.

In a sale he allows a discount of $12\frac{1}{2}\%$ off the marked price.

(b) Calculate the sale price.

(c) Calculate his actual profit, if he sold the coat in the sale.

Unit 9 Speed, distance and time

Aims of the unit

To revise:

1 Time – the 12-hour and 24-hour clocks,

2 Timetables,

3 Distance, time and average speed,

4 Distance/time graphs.

Time – the 12-hour and 24-hour clocks

The time of day can be expressed using either the 12-hour clock or the 24-hour clock. Our normal clocks and watches use the 12-hour clock and we tell morning and afternoon by using a.m. and p.m.

> a.m. = ante meridiem (before midday) = morning
>
> p.m. = post meridiem (after midday) = afternoon/evening

Digital watches and clocks have, these days, enabled us to use the 24-hour clock system. Most travel timetables are now written using the 24-hour clock. It runs from 0000 hours (midnight) through to 2400 hours (midnight again). 7 a.m. (i.e. 7 o'clock in the morning) is written as 0700 ('seven hundred hours'). Midday is 1200 hours and 4.30 p.m. is 1630 hours ('sixteen-thirty hours').

Timetables

Table 1 shows an extract from a bus timetable.

Table 1	Bus A	Bus B	Bus C
Meir Park	0900	1100	1330
Meir Broadway	0907	1106	1337
Normacot	0912	1110	1342
Longton	0918	1115	1348
Fenton	0925	1120	1354
Hanley	0940	1130	1410

We have to be prepared to answer questions of the following type:

(a) How long does it take each of the buses to travel from Meir Park to Hanley?

Bus A takes 40 mins. Bus B takes 30 mins. Bus C takes 40 mins.

(b) Which bus is quickest and by how many minutes?

Bus B is quickest by 10 mins.

(c) If I have an appointment in Hanley at 2 p.m. will I be late if I catch Bus C from Longton?

Yes, I will be 10 mins late arriving at Hanley at 2.10 p.m.

(d) How many minutes between Bus B and Bus C at Normacot?

1110 to 1342 = 2 hours 32 minutes = 152 minutes.

Distance, time and average speed

These three are linked in the following way:

$$\text{Distance travelled} = \text{Average speed} \times \text{Time taken}$$

or $$\text{Average speed} = \frac{\text{Distance travelled}}{\text{Time taken}}$$

or $$\text{Time taken} = \frac{\text{Distance travelled}}{\text{Average speed}}$$

We have to be particularly careful with the units we use. For instance if the average speed is in km/hour, then we must be sure that the distance is in km and the time is in hours.

Example 1 What distance does a car travel at 50 km/hour for 2 hours?

$$\text{Distance} = \text{Speed} \times \text{Time} = 50 \times 2 = 100 \text{ km}$$

Example 2 What is the average speed of a train covering 200 miles in 4 hours?

$$\text{Average speed} = \frac{\text{Distance}}{\text{Time}} = \frac{200}{4} = 50 \text{ miles/hour}$$

Example 3 How long does it take a person to walk 5 km at 10 km/hour?

$$\text{Time} = \frac{\text{Distance}}{\text{Speed}} = \frac{5}{10} = \frac{1}{2} \text{ hour} = 30 \text{ minutes}$$

Distance/time graphs

A graph which shows the connection between distance travelled from a fixed point and the time taken for the journey is often a useful way to represent a journey. Such a graph is known as a distance/time graph.

Example A family travel by car from Birmingham to Manchester, a distance of 150 km. The journey which starts at 0900 hours is shown, with stops made at Wolverhampton and Sandbach (motorway services), in Fig. 1.

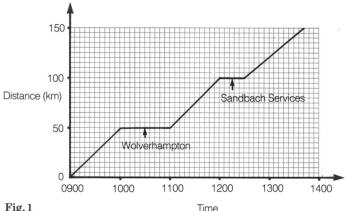

Fig. 1   Time

From the graph find:
(a) The distance travelled in
 (i) the first hour – 50 km.
(ii) the second hour – no distance; 1 hour stop at Wolverhampton.
(b) The total distance travelled – 150 km.
(c) How far were the family from Birmingham at midday? – 100 km.
(d) At what time were they half-way through the journey?
– i.e. 75 km at 1130 hours.
(e) At what time did they arrive in Manchester? – 1345 hours.

NB: When reading from these graphs, and when constructing them, we must take pains to use the scales of both the time and the distance axes very carefully indeed.

Practice – Unit 9   Speed, distance and time

1 Joan leaves her house at 6.45 p.m. to go to the cinema. The journey to the cinema takes 20 minutes. The film lasts for $2\frac{3}{4}$ hours and the return journey home takes 25 minutes. What time did Joan get back to her house?

2 Give these times in the 24-hour clock:
(a) 8.00 p.m.   (b) 9.30 p.m.   (c) 4.45 a.m.
(d) 15 minutes past midnight.   (e) 10 minutes past noon.

3 (a) How far will you travel if you cycle at an average speed of 15 km/hour for 5 hours?
(b) How far will you travel if you drive at 30 mph for 40 minutes?
(c) How long will it take you to cycle 100 km at an average speed of 25 km/hour?
(d) A boy walks for a distance of $22\frac{1}{2}$ km at $4\frac{1}{2}$ km/hour. How long will the journey take?
(e) What is the average speed of a hiker who walks 60 km in 12 hours?
(f) What is the average speed in km/hour of an aircraft which flies 200 km in 30 minutes?

4 Table 2

	London	Birmingham	Cardiff	Edinburgh	Liverpool
Birmingham	110				
Cardiff	155	100			
Edinburgh	380	287	365		
Liverpool	205	94	164	210	
Manchester	190	81	172	210	34

Table 2 gives the distances in miles between six cities in the UK. Use the table to find:
(a) The distance between Cardiff and Manchester.
(b) The distance between London and Birmingham.
(c) The time taken to go from Liverpool to Edinburgh at an average speed of 70 mph.
(d) The average speed of a motorist who takes $5\frac{1}{4}$ hours to go from Manchester to Edinburgh.
(e) The average speed of a motorist who set out from Cardiff at 1325 and arrived in Birmingham at 1555.

5 Table 3 shows an extract from a bus timetable:

Table 3

Longton	1001	then	every	30 minutes	until	1501
Fenton	1008	,,	,,		,,	1508
Stoke	1027	,,	,,		,,	1527
Hanley	1032	,,	,,		,,	1532
Burslem	1036	,,	,,		,,	1536
Tunstall	1055	,,	,,		,,	1555
Kidsgrove	1120	,,	,,		,,	1620
Alsager	1138	,,	,,		,,	1638

(a) How many minutes long is the journey from Longton to Hanley?
(b) How many minutes long is the journey from Tunstall to Alsager?
(c) At what time does the 1131 bus from Longton arrive at Alsager?
(d) A boy arranges to meet his girlfriend in Stoke at 1445. What bus should he take from Longton?
(e) How many minutes will he have to wait for his girlfriend if the bus is 5 minutes late and she is 10 minutes early?

6 In Fig. 2 AB and CD represent roads to and from motorway exits B and C.

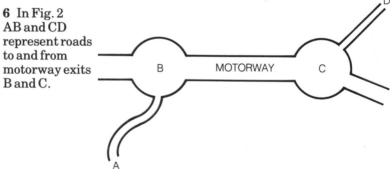

Fig. 2

(a) A car starts at A and travels for half-an-hour from A to B at an average speed of 50 mph. Find the distance from A to B.
(b) The car then travelled on the motorway for 2 hours from B to C a distance of 142 miles. Find the average speed for this part of the journey.
(c) Finally the car left the motorway at C and travelled 69 miles to D at an average speed of 46 mph. Find the time taken in hours to travel from C to D.
(d) Calculate the total distance travelled from A to D.
(e) Calculate the total time taken for the journey from A to D.
(f) Calculate the average speed of the car for the journey from A to D.
7 The travel graph in Fig. 3 records the early morning jog of Mr Jones.

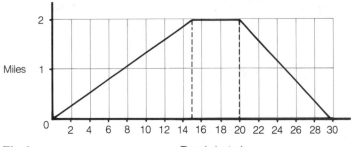

Fig. 3   Time (minutes)

(a) For how long did he rest?
(b) When does he run faster – before or after his rest?
(c) What is his speed for the first part of the journey, in mph?
*8 The graph in Fig. 4 shows the journeys made by a bus and a car. Both travel from Stoke to Coventry and then return to Stoke. The vertical axis shows the distance from Stoke and the horizontal axis shows the time of day.

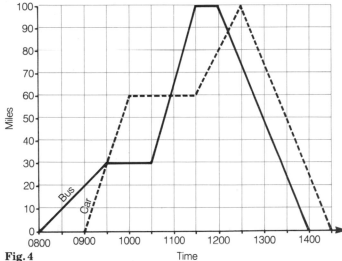

Fig. 4   Time

(a) How many times do the bus and the car pass each other?
(b) How far did the bus travel before it made its first stop?
(c) How far from Coventry was the car when it made its first stop?
(d) How long did the bus wait in Coventry?
(e) What was the speed of the car during the first part of the journey?
(f) How much longer did it take, including stops, for the bus to complete the whole return journey?

Aims of the unit

To revise:
1 Line symmetry,
2 Rotational symmetry,
3 Order of symmetry.

Line symmetry

There are two distinct types of symmetry of which we must be aware, namely LINE symmetry and ROTATIONAL symmetry.

If we fold the letter 'E' along the dotted line (as shown in Fig. 1) the two halves would fit on to each other exactly. The whole shape is said to be SYMMETRICAL and the fold line is called a LINE OF SYMMETRY.

Fig. 1

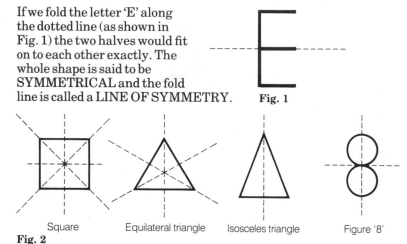

Square Equilateral triangle Isosceles triangle Figure '8'

Fig. 2

Fig. 2 shows some other examples of shapes having line symmetry.

Rotational symmetry

If we rotate a shape about its centre and at some position during the rotation the shape appears not to have moved, then the shape is said to have rotational symmetry.

Example Fig. 3 shows a letter 'N'. If the letter is rotated through 180° (half-turn) about the dot it makes an identical letter 'N' at this point. We therefore say that the letter 'N' has rotational symmetry.

Fig. 3

Fig. 4 shows some other examples of shapes having rotational symmetry.

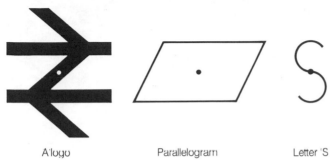

A logo   Parallelogram   Letter 'S'

Fig. 4

Obviously there are many shapes having just line symmetry or just rotational symmetry but there are some shapes which do have both. Fig. 5 shows some examples having both types of symmetry.

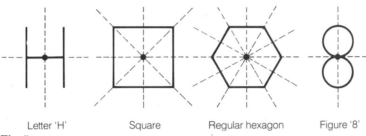

Letter 'H'   Square   Regular hexagon   Figure '8'

Fig. 5

Order of symmetry

The order of rotational symmetry is simply the number of times that a shape fits on to its starting position during its 360° rotation including its starting position. Fig. 6 shows a few examples.

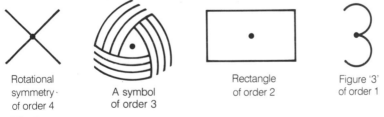

Rotational symmetry of order 4   A symbol of order 3   Rectangle of order 2   Figure '3' of order 1

Fig. 6

Note that all shapes have at least rotational symmetry of order 1. Shapes with rotational symmetry of at least order 2 are sometimes said to have point symmetry.

Practice – Unit 10 Symmetry

1 Draw a ring round any letters below which have two lines of symmetry only:

(a) M A T H S   (b) L I O N

2 Draw a ring round any letters below which have one line of symmetry only:

(a) D I V I D E   (b) A D D I T I O N

3 Draw a ring round any letters below which have rotational symmetry of order 2 (or more):

(a) S I X T E E N   (b) S H A R E

4 Draw the following shapes using the line AB as the axis of symmetry (Fig. 7).

(a)   (b)

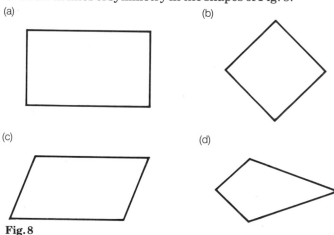

A

B

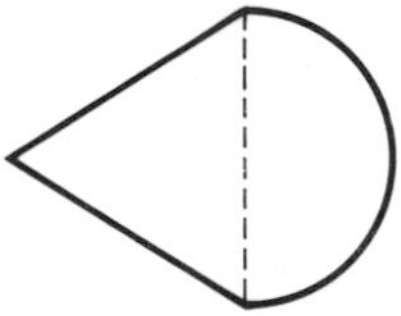

Fig. 7

5 Draw in lines of symmetry in the shapes of Fig. 8.

(a)   (b)

(c)   (d)

Fig. 8

6 Draw the reflection of the pattern of tiles on the left using the dotted line as the axis symmetry (Fig. 9).

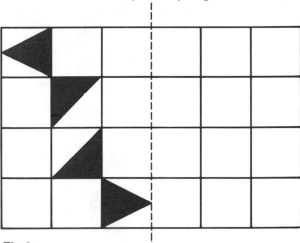

Fig. 9

7 Rotate the shape in Fig. 10 through 90° in a clockwise direction about O.

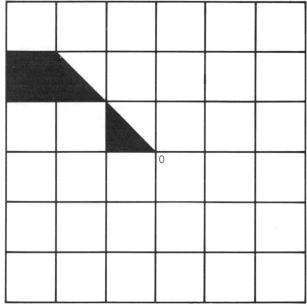

Fig. 10

Now rotate the shape in Fig. 10 through two more 90° clockwise turns about O.

8 State the order of rotation of the symmetrical shapes of Fig. 11.

(a)

(b)

(c)

(d)

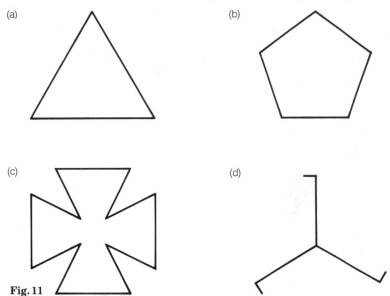

Fig. 11

(e) Name two 4-sided shapes that have rotational symmetry of order 2.

Aims of the unit

To revise nets of simple solid figures.

Nets

If we take a solid shape, such as a cube, and open it out so that it becomes a flat surface (called a NET of the cube), what would it actually look like? Fig. 1 shows very clearly a net (b) of the cube (a).

(a)   (b)

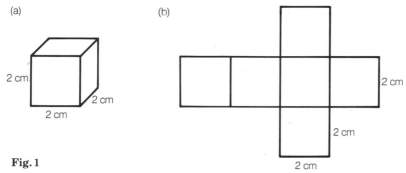

Fig. 1

2 cm

Clearly we can see from this that the cube is made up of 6 surfaces all of the same size. Obviously in the case of a cuboid, opposite sides only of the shape are the same size and so the net is slightly different from that of the cube.

Another net of a familiar shape is shown in Fig. 2.

Fig. 2 Net of an equilateral triangular prism

It is also quite easy to draw the net of an isosceles triangular prism in a similar manner to Fig. 2. With careful practice we can draw a net of many straightforward shapes such as square-based and rectangular-based pyramids etc.

Also, if we wish to make models of solid shape from card, we can draw the net on card, cut out the shape and fold along the edges. The only problem is sticking the shape together. This can be done by including glueing tabs on the net before the cutting-out stage. These are in fact hidden within the finished solid shape.

Practice – Unit 11   Nets

1 Fig. 3 shows part of a net of a cuboid which has dimensions 4 cm by 3 cm by 2 cm.

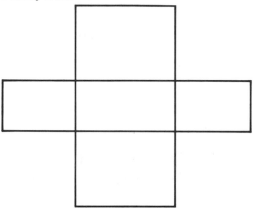

Fig. 3

(a) Complete the net accurately.
(b) Calculate the total surface area of the net.
(c) Calculate the volume of the cuboid.
(If necessary consult Unit 12 for (b) and (c).)
2 Fig. 4 shows the net of a solid.

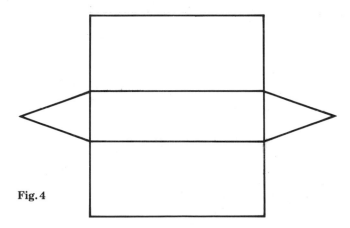

Fig. 4

(a) What is the name given to the solid?
(b) How many faces has the solid?
(c) How many edges has the solid?
*3 Fig. 5 represents the office 'in-tray'. Using a scale of 1 cm to represent 10 cm on the tray, draw an accurate net of the tray.

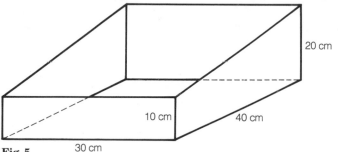

20 cm

10 cm

40 cm

Fig. 5 30 cm

(a) What area of plastic will be needed to make the tray? (See Unit 12.)
(b) If the shape is to be cut out of a rectangular piece, what are the minimum dimensions of the rectangle needed?
(c) What area is wasted? (See Unit 12.)

Aims of the unit

To revise:
1 Perimeter of flat shapes,
2 Area of flat shapes,
3 Volume of solid shapes with uniform cross-section along their length,
4 Surface areas of solid shapes.

Perimeter

The perimeter is the distance all around the edge of a shape. If we consider the shape in Fig. 1, the perimeter will be
6 cm + 3 cm + 3 cm + 1 cm + 3 cm + 2 cm = 18 cm.

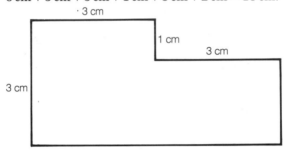

Fig. 1

Distances around the edge of the shape total 18 cm and hence the perimeter is 18 cm.

Remember perimeter is a distance and is measured in cm, m, etc.

Area

We need to know how to find the area of a variety of flat shapes:

K

Area of a square or rectangle = Length × Breadth (Width)

Example What is the area of a rectangular flowerbed measuring 8 m by 4 m (see Fig. 2)?

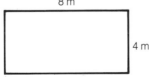

$$\text{Area} = L \times B$$
$$= 8 \times 4 = 32 \, \text{m}^2 \, (\text{Units of}$$
area are square units – in this case square metres, i.e. m^2.)

Fig. 2

$$\text{Area of a triangle} = \frac{\text{Base length} \times \text{Vertical height}}{2}$$

Example 1 What is the area of the triangle shown in Fig. 3?

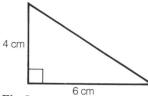

4 cm

6 cm

Fig. 3

$$\text{Area} = \frac{b \times h}{2}$$

$$= \frac{6 \times 4}{2} = \frac{24}{2} = 12 \, \text{cm}^2$$

(This triangle is right-angled, so its vertical height is 4 cm.)

Example 2 Find the area of the triangle shown in Fig. 4.

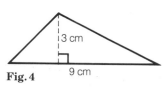

3 cm

9 cm

Fig. 4

$$\text{Area} = \frac{b \times h}{2} = \frac{9 \times 3}{2} = \frac{27}{2}$$

$$= 13\tfrac{1}{2} \, \text{cm}^2$$

(The vertical or PERPENDICULAR HEIGHT of this triangle is 3 cm.)

A PARALLELOGRAM is a 4-sided shape with opposite sides equal in length and parallel to each other.

Area of a parallelogram = Base length × Vertical height

Since a parallelogram is made up of two triangles, its area will be double that of one of the triangles.

Example What is the area of the parallelogram shown in Fig. 5?

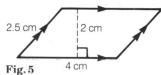

2.5 cm

2 cm

4 cm

Fig. 5

Area = Base × Perpendicular height
   = $4 \times 2 = 8 \, \text{cm}^2$

Remember that it is the vertical height we are using not the *slant* height (2·5 cm in this case).

A TRAPEZIUM is a 4-sided figure with only one pair of parallel sides of lengths a and b, as shown in Fig. 6 with $a = 6$ cm and $b = 8$ cm.

Area of a trapezium = $\tfrac{1}{2}$(Side length a + Side length b) × Perpendicular height

Example Find the area of the trapezium shown in Fig. 6.

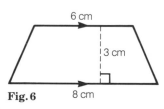

6 cm

3 cm

8 cm

Fig. 6

$$\text{Area} = \tfrac{1}{2}(6 + 8) \times 3$$

$$= \tfrac{1}{2}(14) \times 3$$

$$= 7 \times 3$$

$$= 21 \, \text{cm}^2$$

Area of a circle = πr^2 (where π is a constant number, slightly greater than 3, for any circle – it is usually given as $3\frac{1}{7}$ or $\frac{22}{7}$ or 3·14 or 3·142).

Example Find the area of the circle in Fig. 7. Use $\pi = \frac{22}{7}$.

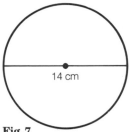

Note that the measurement of 14 cm is the diameter of the circle but in order to find its area we need the radius – remember that radius = half of the diameter = 7 cm (in this case).

Fig. 7

So, Area $= \pi \times r^2 = \pi \times r \times r = \dfrac{22}{7_1} \times 7^1 \times 7 = \dfrac{154}{1} = 154\,\text{cm}^2$

We should, of course, be able to find the perimeter (CIRCUMFERENCE) of a circle. This we can do by using the fact that

Circumference $= 2\pi r$ or πd (since $d = 2r$)

Hence, the circumference of the circle in Fig. 7 will be

$$2 \times \frac{22}{7_1} \times 7^1 = \frac{44}{1} = 44\,\text{cm or } \frac{22}{7_1} \times 14^2 = 44\,\text{cm}$$

To find the area of COMPOSITE SHAPES (shapes made up of two or more shapes), it is usually necessary to find the area of both and add them together or subtract one from the other.

Example 1 What is the total area of the shape in Fig. 8?

Total area of shape
= Area of rectangle +
    Area of triangle
$= (6 \times 3) + (\frac{1}{2} \times 6 \times 5)$
$= 18 + 15 = 33\,\text{cm}^2$

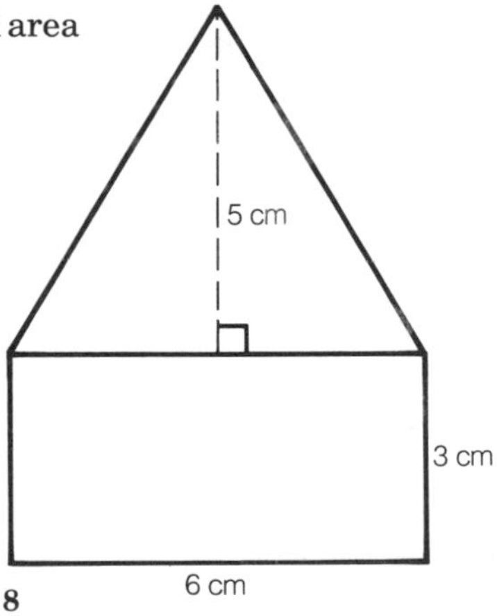

5 cm

3 cm

6 cm

Fig. 8

Example 2 Find the area of the shaded section in Fig. 9 (using a calculator and taking $\pi = 3\cdot142$).

Area of shaded part = Area of triangle − Area of circle

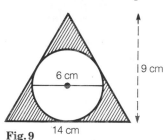

Area of triangle $= \frac{1}{2} \times 14 \times 9 = 63\text{ cm}^2$
Area of circle $= \pi r^2 = 3\cdot142 \times 3 \times 3$
$= 28\cdot278\text{ cm}^2$ (using calculator)
Hence area of shaded part $= 63 - 28\cdot278$
$= 34\cdot72\text{ cm}^2$ (to two decimal places)

6 cm

9 cm

14 cm

Fig. 9

Volume of simple solids

We can find the volume of any solid which has UNIFORM CROSS-SECTION (the shape does not change along its entire length) by finding the area of the cross-section (the end area) and multiplying it by the length of the solid. Such solids of uniform cross-section are called prisms and include such solids as the cube, the cuboid (rectangular box), the triangular prism and the cylinder. Examples of each of these are shown in Figs. 10−13.

Example 1 What is the volume of the cube shown in Fig. 10?

Volume of cube = End area × Length
$= 4 \times 4 \times 4$
$= 64\text{ cm}^3$
(Cubic units used for volume − in this case cubic cm, i.e. cm^3.)

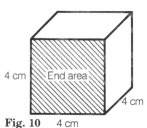

4 cm   End area

4 cm

Fig. 10   4 cm

Example 2 What is the volume of the cuboid shown in Fig. 11?

Volume of cuboid = End area × Length
$= 3 \times 2 \times 6$
$= 36\text{ m}^3$ (cubic metres)

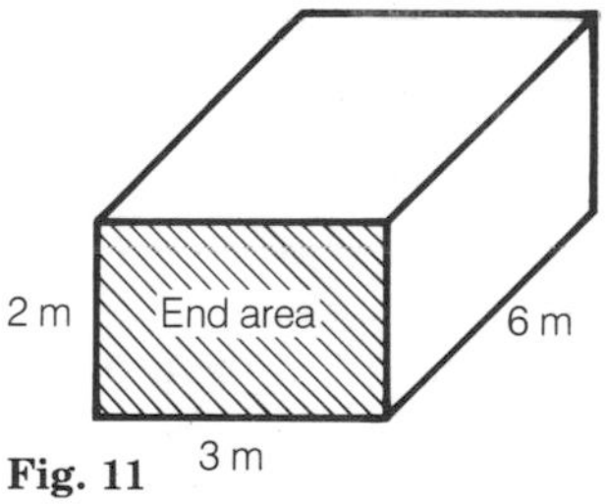

2 m   End area   6 m

Fig. 11   3 m

Example 3 What is the volume of the triangular prism shown in Fig. 12?

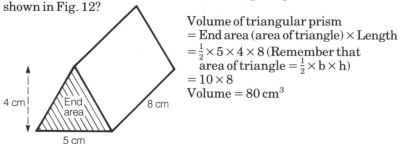

Volume of triangular prism
= End area (area of triangle) × Length
= $\frac{1}{2}$ × 5 × 4 × 8 (Remember that area of triangle = $\frac{1}{2}$ × b × h)
= 10 × 8
Volume = 80 cm^3

Fig. 12

Example 4 What is the volume of the cylinder shown in Fig. 13?

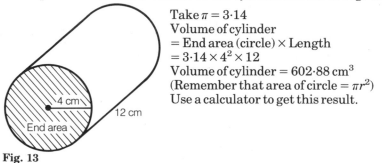

Take π = 3·14
Volume of cylinder
= End area (circle) × Length
= 3·14 × 4^2 × 12
Volume of cylinder = 602·88 cm^3
(Remember that area of circle = πr^2)
Use a calculator to get this result.

Fig. 13

Surface area of simple solids

The surface area of a solid shape is the total area of its surfaces. In the case of a cube which has 6 faces of equal area, we find the area of one face and multiply this by 6 to find the total surface area of the cube (Fig. 14).

Example 1

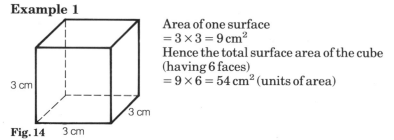

Area of one surface
= 3 × 3 = 9 cm^2
Hence the total surface area of the cube (having 6 faces)
= 9 × 6 = 54 cm^2 (units of area)

Fig. 14   3 cm

In the case of a cuboid the total surface area is made up of the area of the 2 ends, plus the areas of the 2 sides, plus the areas of the top and base (Fig. 15).

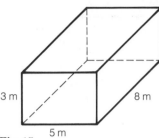

Area of ends $= 3 \times 5 \times 2 = 30\,\text{cm}^2$
Area of sides $= 8 \times 3 \times 2 = 48\,\text{cm}^2$
Area of top and base
$= 8 \times 5 \times 2 = 80\,\text{cm}^2$
Total surface area of cuboid $= 158\,\text{cm}^2$

3 m   8 m

5 m

Fig. 15

Practice – Unit 12 Area, perimeter and volume

1 Fig. 16 represents a square and a rectangle
with the same perimeter.

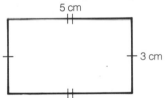

5 cm

3 cm

Fig. 16

(a) Find the perimeter of the rectangle.
(b) Find the side length of the square.
(c) Find the difference in the areas of both shapes.
2 Which triangles in Fig. 17 have the same area as the arrowed one
on the right?

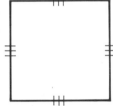

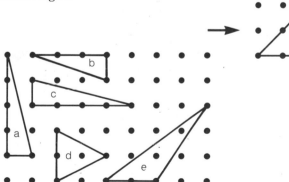

Fig. 17

3 Fig. 18 shows a
piece of wood with a circular
hole cut out. Calculate:
(a) The area of the rectangle.
(b) The area of the hole.
(Take $\pi = 3$; area of a circle $= \pi r^2$.)
(c) The area of wood remaining.

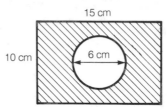

Fig. 18

4 Fig. 19 shows a garden plot 40 m by 30 m.

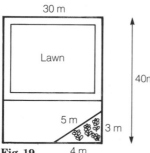

Fig. 19

The square lawn has a path 1 m wide
surrounding it. It also has a
triangular flowerbed with sides 3 m by
4 m by 5 m.
(a) Find the area of the garden plot.
(b) Find the length of the lawn.
(c) Find the area of the concrete path.
(d) Find the area of the triangular
flowerbed.

5 A wheel has 8 spokes (Fig. 20). Each spoke is 20 cm long.
Taking π to be 3 calculate:
(a) Area of the wheel (area $= \pi \times r^2$).
(b) Circumference of the wheel
(circumference $= \pi \times$ diameter).
(c) How far will the wheel travel,
in metres, if it goes round
10 times?

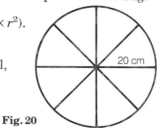

Fig. 20

6 Fig. 21 shows the plan of a garden to be completely fenced in.
(a) Find the total perimeter
of the garden.
(b) Fencing sections can be
bought in lengths of 2 m
at £8 per section.
How much would it cost to fence
in the whole garden?
(c) Calculate the area of the garden. **Fig. 21**

7 Small boxes measuring 3 cm by 2 cm by 2 cm are packed into a
larger carton measuring 10 cm by 10 cm by 15 cm (see Fig. 22).

(a) Calculate the volume of the large carton.
(b) Calculate the volume of the small box.
(c) How many boxes will fit into the carton?

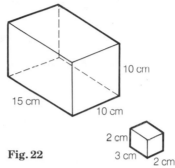

Fig. 22

*8 Calculate the area of the shapes in Fig. 23.

(a)

(b)

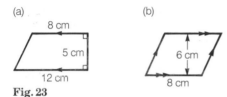

Fig. 23

*9 A cylindrical tin of beans has a diameter of 10 cm and a height of 15 cm. Calculate:
(a) The volume of the tin.
(b) The area of the label around the tin.
(c) The area of tin to make the can. (Take $\pi = 3.142$; Volume $= \pi r^2 h$.)
*10 A concrete gully is made to the specifications of Fig. 24. If $\pi = 3.142$, work out:
(a) The end area of cross-section.
(b) The volume of the gully.
(c) The mass of the gully if concrete has a density of 10 g/cm³.
Answer in kg. Answers correct to 2 dec. pl.

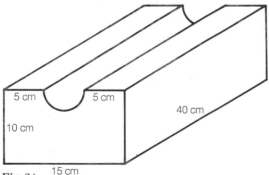

Fig. 24

Aims of the unit

To revise:
1 Collecting, recording and sorting information,
2 The visual representation of data,
3 Averages – mean, median, mode,
4 Simple probability.

STATISTICS is the name given to the science of collecting, studying and analysing facts (called DATA).

Collecting data

Data can be collected in the following ways, for example:
(a) Questioning people and recording their replies – on what are known as Questionnaires.
(b) Observation – by watching and recording the information.
(c) Carrying out experimental work and recording the results.
(d) Finding data published in books and magazines which has been collected by other people.

Recording and sorting information

Having recorded the data in unsorted fashion we then have to sort it into a form which will be useful to us. By far the best way of recording information is to make a FREQUENCY DISTRIBUTION TABLE which enables us to put the unsorted results into some type of order by using a TALLY CHART.

Example The marks of a group of 40 pupils scored out of a maximum 10 in a maths test are shown in the table below:

```
8  7 5 4 4 5 4 6 5 6
6  7 5 3 2 5 7 1 4 7
6  6 8 7 8 7 3 5 7 9
7 10 6 6 3 7 7 4 4 2
```

In this form it is very difficult to make much use of this information; but if we draw up a tally chart and frequency table, as shown in Table 1, then the information becomes far more meaningful.

Table 1 Marks in maths test (out of 10)

Mark	Tallies	Frequency
0		0
1	I	1
2	II	2
3	III	3
4	IIII I	6
5	IIII I	6
6	IIII II	7
7	IIII IIII	10
8	III	3
9	I	1
10	I	1
		Total 40

From the original list of scores one stroke is marked each time the score occurs. Every fifth stroke is drawn across the previous four, giving blocks of five. It is a good idea to cross out the numbers on the list as they are tallied in order to avoid any confusion. It is also important to check the total of the frequencies to see that it is the same as the number of scores, i.e. 40. If a number has been missed or recorded twice then this can be put right by a re-tally. It is vital that we concentrate very hard indeed while performing the tally.

Table 2 is a horizontal version of the same frequency table.

Table 2

Mark	0	1	2	3	4	5	6	7	8	9	10
Frequency	0	1	2	3	6	6	7	10	3	1	1

(Total frequency = 40)

This is an example of collecting and tabulating (UNGROUPED DATA), so called because we are only concerned about a small range of values, i.e. from 1 to 10.

If, however, the marks had been given as marks out of 100, then we would have had to produce a tally chart of enormous size. In order to avoid this problem, we then have to put the marks into classes or groups (known as GROUPED DATA), such as 1 to 10, 11 to 20 etc.

When the marks are grouped together with 10 marks per 'class interval' then the distribution table only needs 10 classes, hence needing only 10 lines to represent it.

Example 40 pupils recorded the following marks in a maths examination (out of a possible 100)

```
72 66 54 63 29 63 27 63 77 65
52 69 63 33  8 63 32  9 56 35
92 83 26 22 46 36 42 51 55 59
47 55 42 11 73 45 56 37 54 44
```

Table 3

Class interval marks	Tally marks	Frequency
0–9	II	2
10–19	I	1
20–29	IIII	4
30–39	IИІ	5
40–49	IИІ I	6
50–59	IИІ IIII	9
60–69	IИІ III	8
70–79	III	3
80–89	I	1
90–→	I	1
		40 Total

The frequency distribution, Table 4, of this data could then be written horizontally as follows:

Table 4

Marks (class intervals)

0–9	10–19	20–29	30–39	40–49	50–59	60–69	70–79	80–89	90→
2	1	4	5	6	9	8	3	1	1

Frequency (number of pupils)

Although we lose some of the detail of the distribution by 'grouping' the data, it is obviously much easier for us to deal with in this way.

Practice – Unit 13 Tally charts

1 By using a tally chart, draw up a frequency distribution table for the goals scored by a hockey team over a period of 30 matches.

6 1 6 2 6 3 6 5 3 3

1 3 7 3 6 0 5 2 6 2

7 8 5 9 7 7 9 8 6 8

*2 The data below show the masses of 50 people attending a 'weight-watchers' class. Using class intervals 50–54, 55–59 etc., construct a tally chart and frequency distribution table to represent the data.

77 54 66 63 74 70 61 67 74 83

71 52 76 64 67 68 71 68 65 62

69 73 64 63 68 79 65 64 72 88

71 79 61 76 66 86 74 84 81 70

75 59 63 66 71 57 68 55 57 67

Visual representation of data

Having collected the data it is then very important to present it in a form which can be easily noticed and understood. The most common ways of showing information are the PICTOGRAM, the BAR GRAPH, the PIE CHART and the HISTOGRAM.

The *pictogram* (or ideograph or pictograph) – this is a simple way of showing facts using little drawings or symbols. In order to produce a good pictogram the following points should be observed:

1 Use small sensible symbols to represent the data,
2 Make the symbols simple and clear,
3 Produce a scale to show exactly what each symbol represents,
4 Make sure that all full symbols are of the same size,
5 Space the symbols equally,
6 Draw all symbols for one section of data in a straight line (either vertical or horizontal),
7 Show data of less than scale size by a part of the symbol.

It is usual to draw symbols similar to the data to be represented, either vertically or horizontally, depending on the data involved. Typical symbols can include such things as: cars, boats, houses or cakes.

Example Study the pictogram in Fig. 1 of cake sales in a shop over a one-week period and write down how many cakes were sold on:
(a) Wednesday.   (b) Friday.   (c) Sunday (and why?).

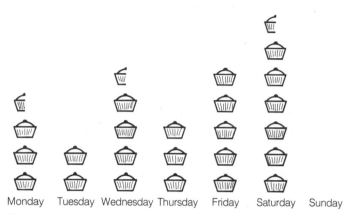

Fig. 1

(a) $4\frac{1}{2}$ cakes represents 450 cakes sold on Wednesday.
(b) 5 cakes represents 500 cakes sold on Friday.
(c) No cakes sold – shop obviously not open on Sunday.

The *bar graph* (or bar chart) – this is the most popular method of displaying statistical data and is probably the easiest of all the different methods we are considering. Once again the information on a bar graph may be displayed either vertically (sometimes called a column graph) or horizontally – both ways being equally acceptable. The bar chart itself is made up of a series of bars. Each bar is usually separated by a space from other bars. Each bar shows a section of the data to be represented. The length (horizontal) or height (if vertical) of each bar depends on the size of the section of the data it represents.

Note that all the bars are the same width. Spaces are also the same width as each other, although not necessarily of the same width as the bars.

Example Of 36 teachers in a school, 9 teach Maths, 7 teach English, 6 teach Humanities, 4 teach Science, 5 teach Art and Crafts, 3 teach Modern Languages and 2 teach PE/Games. Draw a vertical and horizontal bar chart to represent this information (Fig. 2).

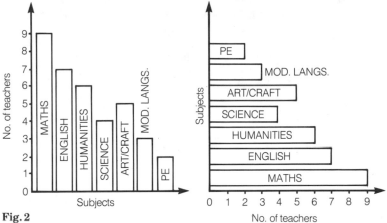

Fig. 2

The *pie graph* (or pie chart) – this method involves drawing a circle which represents a 'pie' of which each slice represents an item. The size of the slice depends on the size of the item. The angle at the centre of a circle is 360°.

Example The pie chart shown in Fig. 3 shows how a boy spends his £4 pocket money each week. How much does he spend on (a) discos, (b) magazines/comics, (c) football match, (d) savings?

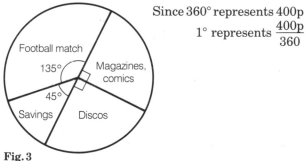

Since 360° represents 400p

$$1° \text{ represents } \frac{400p}{360}$$

Fig. 3

(a) Discos will cost $90 \times \dfrac{400p}{360} = 100p = £1$.

(b) Magazine/comics will cost $90 \times \dfrac{400p}{360} = 100p = £1$.

(c) Football match costs $135 \times \dfrac{400p}{360} = 150p = £1.50$.

(d) He saves $45 \times \dfrac{400p}{360} = 50p$.

The *histogram* is somewhat similar to a vertical bar chart or column graph, except that the *area* of each bar represents the data, rather than the length of the bar. The area of a bar is given by Height × Width with each bar being in contact with the bars on either side. This type of graph is very useful for representing information in which class intervals are not equal in size. However, we shall only consider data of equal class sizes.

Example The histogram in Fig. 4 represents the frequency distribution of the mass (in kg) of 80 pupils. Draw a frequency distribution table of the graph and say which class interval has (a) the most members in it, (b) the fewest members in it.

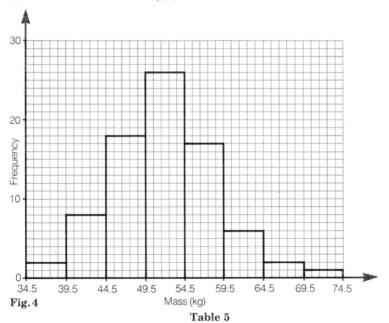

Fig. 4

Table 5

Mass (kg)	35–39	40–44	45–49	50–54	55–59	60–64	65–69	70–74
Numbers of pupils	2	8	18	26	17	6	2	1

(a) The 50–54 kg group has most members in it.
(b) The 70-74 kg group has fewest members in it.

NB: Although the pictogram may be eye-catching its accuracy is somewhat limited. The pie chart is a useful way of comparing one element of a population to the total population. The bar chart/ histogram is a suitable way of directly comparing elements of a population with each other.

Practice – Unit 13 Pictograms, bar charts, pie graphs and histograms

1 Look at the bar chart in Fig. 5 which shows the monthly spending of a typical family. Use the bar chart to find:
(a) How much is spent on heating.
(b) How much is spent on food.
(c) How much is spent altogether in a month.

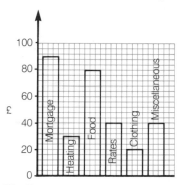

Fig. 5

2 A second-hand-car salesman has 90 cars on his forecourt for sale as shown in the pie chart in Fig. 6:
(a) Calculate how many cars he has for sale of each type.
(b) What fraction of the total number of cars are Nissan?
(c) What percentage of the total number of cars are Peugeot/Talbot?

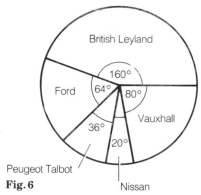

Fig. 6

3 A display of packaged cheeses in a particular 'chill' display in a shop is counted and the number of packets of each type is shown in the pictogram (Fig. 7).

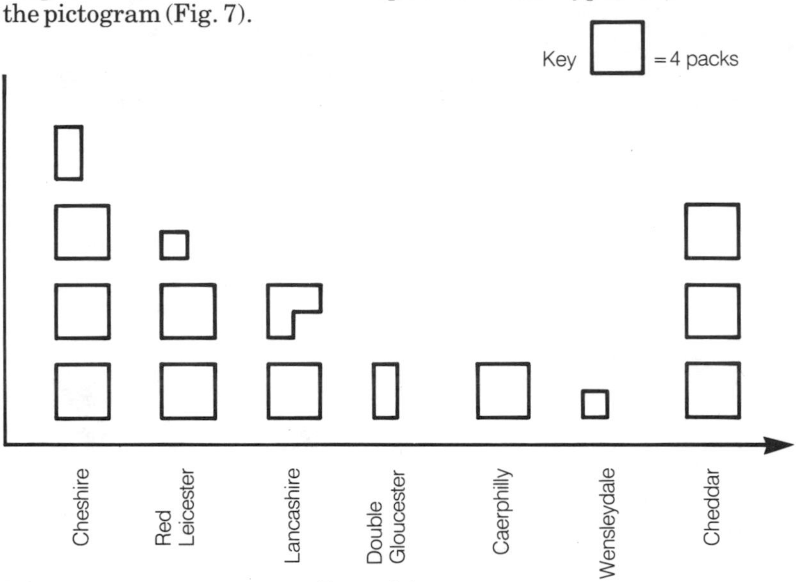

Key ☐ = 4 packs

Fig. 7   Types of cheese

(a) Which cheese is most displayed?

(b) Which cheese is least displayed?

(c) How many packs of Lancashire are displayed?

(d) How many packs of cheese are displayed altogether?

(e) A customer buys 3 packs of Cheshire, 2 packs of Lancashire, 1 pack of Caerphilly and 4 packs of Cheddar. Redraw the pictograph to show the packs remaining on display.

*4 In a survey, 30 girls were asked how many valentine cards they had received. The results are shown in the histogram (Fig. 8).

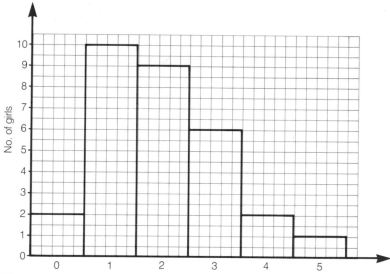

Fig. 8   No. of cards received

(a) How many girls received more than 2 valentine cards?
(b) Find the total number of valentine cards received by the 30 pupils.
(c) What fraction of the girls received 6 cards?
(d) How many girls received 4 or more cards and what percentage of the total number of girls is this?

Averages

When we collect data and put it into an orderly form in the shape of a frequency distribution, one value which often stands out is the AVERAGE of the group. In statistics, we use three 'averages'. These are:

1 The arithmetic mean (often called simply – the mean),
2 The median,
3 The mode.

1 The ARITHMETIC MEAN is found by adding all the values together and then dividing this amount by the number of values taken.

Example Find the mean of 60, 33, 26, 50, 48.

$$\text{Mean} = \frac{60 + 33 + 26 + 50 + 48}{5} = \frac{217}{5} = 43.4$$

When faced with finding the mean of a large amount of data it is far simpler to use a frequency distribution of the data. In the tally chart we construct an extra column to the right containing the values multiplied by the frequencies. Adding the frequencies gives us the number of values in the data and adding the numbers in the new column gives us the data total. The mean is then calculated as before, by dividing the data total by the total of the frequencies.

Example Find the mean of the numbers given:

9 1 7 7 4 5 7 1 1 8

9 6 9 5 6 3 10 1 8 4

4 6 6 2 1 8 0 3 8 1

8 5 9 6 3 3 4 4 6 2

This is put into a tally chart/frequency distribution form as shown in Table 6.

Table 6

Value	Tally	Frequency	Value × Frequency
0	I	1	$0 \times 1 = 0$
1	IIIII I	6	$1 \times 6 = 6$
2	II	2	$2 \times 2 = 4$
3	IIII	4	$3 \times 4 = 12$
4	IIIII	5	$4 \times 5 = 20$
5	III	3	$5 \times 3 = 15$
6	IIIII I	6	$6 \times 6 = 36$
7	III	3	$7 \times 3 = 21$
8	IIIII	5	$8 \times 5 = 40$
9	IIII	4	$9 \times 4 = 36$
10	I	1	$10 \times 1 = 10$
	Total frequency 40		Total of data 200

$$\text{Mean} = \frac{\text{Total of data}}{\text{Frequency total}} = \frac{200}{40} = 5$$

2 The MEDIAN of a distribution is the 'middle' value when the data is arranged in numerical order. When there is an odd number of values, the median is simply the middle value. However, when there is an even number of values, the median is found by taking the mean of the 'middle two' values.

Example 1 (Odd number of values) Find the median of:

6, 2, 3, 1, 5, 4, 3, 3, 1

In numerical order: 1, 1, 2, 3, $\underline{3}$, 3, 4, 5, 6

Nine numbers, hence the median is the fifth number, i.e. median = 3

Example 2 (Even number of values) Find the median of:

3, 4, 2, 3, 5, 2, 4, 6, 5, 1

In numerical order: 1, 2, 2, 3, $\underline{3}$, $\underline{4}$, 4, 5, 5, 6

Ten numbers, hence the median is the mean of the 'middle two' numbers, i.e. median $= \dfrac{3+4}{2} = 3 \cdot 5$

Once again we can use a tally chart (Table 7) to form a frequency distribution for large amounts of data.

Example Find the median of the following data:

54, 51, 54, 52, 53, 55, 54, 55, 53, 51, 52, 54,
53, 52, 53, 56, 54, 53, 55, 51, 54, 53, 53, 52

Table 7

Value	Tally	Frequency
51	III	3
52	IIII	4
53	IЖ II	7 ← 12th and 13th values
54	IЖ I	6 lie here
55	III	3
56	I	1
	Frequency total = 24	

We look in Table 7 for the position of the 12th and 13th values since they are the 'middle two' values of the 24 total frequency. On examination, we find that both the 12th and 13th values of the distribution are 53 and so the median is 53.

3 The MODE of a set of data is the value which occurs most frequently, i.e. the value which has the greatest frequency. It is probably the easiest of the three 'averages' to find.

Example Find the mode of the following set of numbers:

3, 4, 1, 3, 5, 2, 1, 5, 1, 2

1 occurs most often (i.e. 3 times)
hence the mode of this distribution = 1.

As before, we can use a tally chart (Table 8) to form a frequency distribution from which the mode is immediately available.

6, 6, 3, 6, 8, 7, 1, 7, 4, 9, 5, 8, 4, 2, 5, 4, 3, 7,
5, 7, 5, 5, 2, 6, 6, 4, 8, 5, 6, 3, 6, 4, 8, 5, 3, 7,
4, 7, 6, 9, 6, 6, 2, 4, 7, 4, 1, 6, 8, 5, 5, 3, 7, 9

Table 8

Value	Tally	Frequency
1	II	2
2	III	3
3	⦀ЖІ	5
4	ЖІ III	8
5	ЖІ IIII	9
6	ЖІ ЖІ 1	11 ← greatest frequency
7	ЖІ III	8
8	ЖІ	5
9	III	3

Since the value 6 has the greatest frequency, namely 11, the modal
value is 6.

Practice – Unit 13 Means, medians and mode

1 A football team scored the following number of goals in 10 matches:

1, 0, 2, 5, 1, 2, 0, 2, 3, 2

Find (a) the mean score, (b) the median score, (c) the modal score.
*2 The following numbers are the number of words in each line of a
page chosen from the novel *Treasure Island* by Robert Louis
Stevenson.

13, 8, 11, 2, 9, 12, 11, 12, 10, 9, 12, 10, 7, 3
12, 12, 12, 12, 8, 10, 1, 12, 9, 13, 10, 10, 1, 2,
13, 12, 1, 11, 12, 9, 6, 10, 8, 13, 1, 7, 7, 7

Construct a frequency table and use it to find:
(a) The mean number of words per line.
(b) The median number of words per line.
(c) The modal number of words per line.

Probability

Probability is a measure of the chance (or likelihood) of an event
happening. The scale of measurement of probability of an event
ranges from zero to one and is measured either in fractional or
decimal form, i.e. $\frac{1}{2}$ or 0·5 etc.

ZERO represents IMPOSSIBILITY and ONE represents CERTAINTY

Hence $0 \leqslant P(\text{event}) \leqslant 1$

Finding probabilities

There are two ways of finding the probability of an event such as spinning a coin or throwing a die: (a) by experiment and (b) by calculation.

(a) Probability by experiment. By repeating the experiment (i.e. tossing the coin many times) the probability fraction may be calculated as follows:

$$\frac{\text{Number of times event occurred during experiment}}{\text{Total number of trials performed in experiment}}$$

In the case of P(a tossed coin will land tails) the fraction will be

$$\frac{\text{Number of tails recorded}}{\text{Total number of tosses}}$$

In the case of P(throwing a '5' with a die) the fraction will be

$$\frac{\text{Number of '5's recorded}}{\text{Total number of throws}}$$

We must remember that the probability of an event worked out by experiment is only an estimate of the actual probability of that event. If an experiment consists of a large number of trials, then the chances of the experiment producing good estimates of the actual probabilities is far greater. In fact the larger the number of trials the better the estimates will become.

(b) Probability by calculation (expected probability). When all the possible results are equally likely we can calculate the EXPECTED PROBABILITY that an event will occur from the formula:

$$P(\text{event occuring}) = \frac{\text{Number of times event happens}}{\text{Total number of happenings}}$$

Example 1 What is the probability of drawing a 'king' from a pack of cards?

Number of times event can happen (i.e. number of 'kings' in pack) = 4

Total number of possible happenings (i.e. total number of cards in pack) = 52

and so $P(\text{king}) = \dfrac{4}{52} = \dfrac{1}{13}$

Example 2 A bag contains 20 wine gums: 8 yellow, 6 green, 3 red, 2 black and 1 orange. Find the probability of pulling a black wine gum out of the bag at the first attempt.

Number of black gums = 2
Total number of gums in bag = 20
and so P(black wine gum) $= \dfrac{2}{20} = \dfrac{1}{10}$

Expected number

The expected number = Probability of success $\times$
Number of attempts

Example If I toss a coin 50 times, how many times would I expect the coin to land tails?

P(tail) $= \frac{1}{2}$, number of attempts = 50
Expected number of tails $= \frac{1}{2} \times 50 = 25$
I would expect 25 'tails' out of 50 tosses

NB: The expected number will only be an estimate because, in an actual experiment, we could 'reasonably' expect any result between 20 and 30 'tails' – it could of course be even more or less. Once again, the greater the number of trials the more likely the number would be closer to the expected number.

Complementary events

We consider here the probability of an event *not happening*.
Since we know that an event will either take place or not take place, the sum of the probabilities must be 1 i.e. $P(A) + P(\text{not } A) = 1$.
   Rearranging the above equation, $P(\text{not } A) = 1 - P(A)$. So, to find the probability of an event not happening, it is quite often best to find the probability that the event will happen and subtract this answer from 1.

Example A bag contains 5 blue sweets, 3 red sweets and 2 yellow sweets. What is the probability of not taking out a red sweet at the first attempt?

P(red sweet) $= \dfrac{3}{10}$

$\therefore P$(not a red sweet) $= 1 - \dfrac{3}{10} = \dfrac{7}{10}$

Practice – Unit 13 Simple probability

*1 If a letter is taken at random from the word 'STATISTICS', what is the probability that it will be:
(a) an E, (b) a T, (c) a vowel, (d) an I, (e) a consonant?
*2 A bag contains 10 marbles. If there are 5 green, 3 yellow and 2 blue marbles, what is the probability that if one is taken out of the bag it will be:
(a) blue, (b) yellow, (c) green, (d) red?
*3 When a die is thrown, what is $P(5)$ and what is $P(\text{not }5)$?
*4 A box contains 4 white, 3 yellow, 2 blue and 1 red pencils. One is taken out at random. Find:
(a) $P(\text{white})$.   (b) $P(\text{not yellow})$.
(c) $P(\text{red})$.   (d) $P(\text{not blue})$.

Sample spaces (possibility diagrams)

A list or table of all the possible outcomes which follow from a particular course of action is called a SAMPLE SPACE.

Example 1 The sample space when a normal six-sided die is thrown is:

   1 2 3 4 5 6

Example 2 The sample space when a coin is tossed is:

   head   tail

Combined events

Sample spaces are even more useful when the probabilities of events which result from a combination of two (or more) separate happenings, are required.
   A COMPOSITE sample space is formed by combining the sample spaces for each separate happening. Consider, for example, the composite sample space in Table 9 for the total scores of two dice being thrown at the same time.

Table 9   Die 2

+	1	2	3	4	5	6
1	2	3	4	5	6	7
2	3	4	5	6	7	8
3	4	5	6	7	8	9
4	5	6	7	8	9	10
5	6	7	8	9	10	11
6	7	8	9	10	11	12

Die 1 (rows 1–6)

We can use the composite sample space to find the probabilities of the following: (a) a score of 8; (b) a score greater than 9; (c) a score less than or equal to 5; (d) a score of not 7; (e) a score of 13.

Since there are 36 possible outcomes in the sample space:

(a) $P(\text{score of } 8) = \dfrac{5}{36}$.

(b) $P(\text{score} > 9) = \dfrac{6}{36} = \dfrac{1}{6}$.

(c) $P(\text{score} \leqslant 5) = \dfrac{10}{36} = \dfrac{5}{18}$.

(d) $P(\text{score} \neq 7) = 1 - P(\text{score of } 7) = 1 - \dfrac{6}{36} = \dfrac{30}{36} = \dfrac{5}{6}$.

(e) $P(\text{score of } 13) = 0$ (since 13 does not appear in the sample space).

NB: The sample space makes the problem much easier to visualize and helps us considerably. So long as the events are separate happenings then the sample space makes the calculation quite easy.

Probability tree diagrams

K Another very useful method of illustrating independent events is by drawing a PROBABILITY TREE. With such a picture before us, the situation is much clearer.

Example Suppose we have a bag containing 7 black balls and 3 white balls. One ball is taken out at random and not replaced; then a second is removed. We can draw a 'tree' diagram to represent the situation, as shown in Fig. 9.

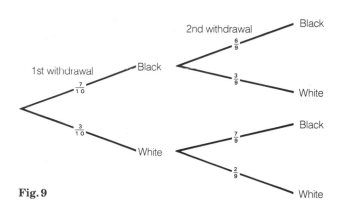

Fig. 9

Now since the two withdrawals are independent events we can use the rule: $P(\text{A and B}) = P(\text{A}) \times P(\text{B})$ and, using our 'tree', we can now find the probabilities of such happenings as

$$P(\text{black and black}) = \frac{7}{10} \times \frac{6}{9} = \frac{7}{15}$$

$$P(\text{black and white}) = \frac{7}{10} \times \frac{3}{9} = \frac{7}{30}$$

$$P(\text{white and black}) = \frac{3}{10} \times \frac{7}{9} = \frac{7}{30}$$

$$P(\text{white and white}) = \frac{3}{10} \times \frac{2}{9} = \frac{1}{15}$$

NB: When added together, probabilities of all these happenings total 1 (i.e. certainty). It is worth noting that probabilities on adjacent branches of the tree always total 1 (which is a useful check for errors). We could of course extend the tree to take account of further withdrawals.

Practice – Unit 13   Sample spaces and probability trees

*1 Draw a sample space diagram to show the results when a die and a coin are thrown together. Use the diagram to help you to find the probability of:
(a) A tail and a 3 showing.
(b) A head and a 5 showing.
(c) A tail and a 7 showing.
*2 A box contains 3 yellow and 5 blue counters. One is drawn out at random and not replaced and a second draw is then made. Draw a tree diagram to show the situation. Use your tree diagram to help you to find:
(a) $P(\text{yellow first and yellow second})$.
(b) $P(\text{yellow first and blue second})$.
(c) $P(\text{blue first and yellow second})$.
(d) $P(\text{blue first and blue second})$.

Unit 14 Algebra

Aims of the unit

To revise:
1 The use of letters and the making of algebraic statements,
2 Basic arithmetic processes expressed algebraically,
3 Substitution of numbers for letters,
4 Construction and solution of simple equations,
5 Transformation of simple formulae,
6 Use of brackets and simple factorization,
7 Positive and negative indices.

Algebraic statements

K ▶ Algebraic statements are expressions constructed from algebraic information rather than arithmetic information. In order to complete them successfully it is a good idea to think in terms of actual numbers while trying to build up the statement.

Example 1 I have 35p in my pocket and I spend xp, how much will I have left?

If x was, say, 15p then I would have left $(35 - 15)\text{p} = 20\text{p}$. So, in general terms, if I spend xp I will have $35 - x$p left.

Example 2 I think of a number, double it and then add 5. What will the new number be?

Suppose the number is x at the start. If I double it, it becomes $2x$. Then adding 5 gives $2x + 5$ (this is the new number).

Basic arithmetic process in algebra

As is the case in arithmetic we can similarly use the operations of addition, subtraction, multiplication and division in algebra. We must, however, remember the basic rules involved. So, for addition and subtraction we can only collect 'like terms'.

Example 1 Simplify $4x + 3x + 2x + x = 10x$.

Example 2 Simplify $3a + 8b + 6a + 4b$.

In this example we have to collect the as and bs separately.

$$3a + 8b + 6a + 4b = 9a + 12b$$

Example 3 Simplify $7a - 3b - 11a + 6c + 9b - 8c$.

In this example we have some subtractions to perform (see Directed number, Unit 3).

$$\text{Hence } 7a - 3b - 11a + 6c + 9b - 8c = -4a + 6b - 2c$$

Example 4 $5a^3 - 2a^3 = 3a^3$.

In the case of multiplication and division we must also refer to Unit 3 – Multiplying and dividing directed numbers.

Example 1 $3 \times a = 3a$.
Example 2 $-a \times c = -ac$.
Example 3 $27x \div 3 = 9x$.
Example 4 $-24yz \div -8z = 3y$.

Substitution of numbers for letters

A letter used in an algebraic expression may have a numerical value.

Example 1 If $a = 4$ find the value of $5a + 3$.

In this case $5a + 3 = (5 \times 4) + 3 = 20 + 3 = 23$

Example 2 If necessary, consult Unit 3 (Directed number) once again.

If $a = -2$, $b = 3$, $c = -4$, find the value of $4a + 5b - c$
$= (4 \times -2) + (5 \times 3) - (-4)$
$= -8 + 15 + 4 = 11$

Example 3 Problem type. The cost (P pence) of framing a picture depends on its length (L cm) and its breadth (B cm). If $P = 2L + 3B$, find the cost if $L = 60$ cm and $B = 40$ cm.

Then $P = (2 \times 60) + (3 \times 40) = 120 + 120 = 240p = £2.40$
Hence the cost of framing the picture will be £2.40

Construction and solution of simple equations

We can build up an ALGEBRAIC EQUATION in exactly the same way as we formed statements in algebra.

Example I think of a number, I treble it and subtract 8. The result is 4. Form an equation and solve it to find the number I thought of. Let the number I thought of be x.

Hence the equation will be $3x - 8 = 4$

Remember that to solve the equation the aim is to have the xs on one side of the equation and the numbers on the other side. In this case we have to add 8 to both sides giving $3x = 12$ then dividing both sides by 3 gives $x = 4$ (which is the number I first thought of).

Following are two more examples of equation solving.

Example 1 Solve $7x - 7 = 3x + 9$.

We want xs on the left-hand side, i.e. subtract $3x$ from both sides, and we want the numbers on the right-hand side, i.e. add 7 to both sides.

Thus $7x - 7 = 3x + 9$
becomes $4x = 16$ then dividing both sides by 4
gives $x = 4$

Example 2 Solve $\frac{2}{3}y = 12$.

Multiply both sides by 3 to remove the denominator giving $2y = 36$; then divide both sides by 2 to find y.
Hence $y = 18$.

Transformation of formulae (change of subject)

K ▶ A formula is basically an equation which gives a general rule for a particular problem. See the following examples.

Example 1 For the circumference of a circle (C) we use the formula $C = 2\pi r$ (where r is the radius of the circle).

It is of course possible to TRANSFORM (change the subject of) the formula $C = 2\pi r$ by dividing both sides by 2π.

Hence $\dfrac{C}{2\pi} = r$

The radius r is now the subject of the formula.

Example 2 Simple interest is found by using the formula $I = \dfrac{PRT}{100}$.

Rearrange the formula to make T the subject of the formula.

Multiplying both sides of $I = \dfrac{PRT}{100}$ by 100 gives

$100I = PRT$

In order to isolate T we divide both sides by PR giving

$\dfrac{100I}{PR} = T$

Removal of brackets

K ▶ In order to remove brackets we have to multiply each term inside the bracket by the number outside the bracket.

Example 1 $5(y + 2) = 5 \times y + 5 \times 2 = 5y + 10$.

Example 2 $-3(2a - 4b) = -3 \times 2a + -3 \times -4b = -6a + 12b$.

Example 3 $-(2x - y)$ is treated as $-1(2x - y)$, i.e. $-2x + y$.

Example 4 Remove the brackets and simplify the following expression $3(2a + 3b) - 2(a - b)$.

$= 6a + 9b - 2a + 2b$ collect like terms
$= 4a + 11b$ (in simplest form).

Basic factorization

In factorizing an expression, we are effectively reversing the process of bracket removal by introducing brackets into an expression. We perform this operation by looking for the highest common factors within the expression.

Example 1 Factorize $3x + 3y = 3(x + y)$

$$\uparrow$$

Common factor

NB: You can check your factorizing by removing the brackets.

Example 2 Factorize $4a - 6b = 2(2a - 3b)$.

$$\uparrow$$

Common factor

Example 3 Factorize $-3m - 12 = -3(m + 4)$.

$$\uparrow$$

Common factor

Example 4 Factorize $x + x^2 = x(1 + x)$.

$$\uparrow$$

Common factor

If you are unsure about this, consult the next section – Indices.

Powers and indices

Multiplication of powers. Powers of the *same* number are multiplied together by adding the indices, i.e. $a^m \times a^n = a^{(m + n)}$.

Example $d^3 \times d^2 = d^{(3 + 2)} = d^5$.

Division of powers. Powers of the *same* number may be divided by subtracting the indices, i.e. $a^m \div a^n = a^{(m - n)}$.

Example $m^6 \div m^2 = m^{(6 - 2)} = m^4$.

Negative indices. We know that a^2 means $a \times a$, but what does a^{-2} mean? We have here a negative power and $a^{-2} = \dfrac{1}{a^2}$. In general terms:

$$a^{-m} = \frac{1}{a^m}$$

Remember also that for any value of a, $a^0 = 1$.

Now consider the following examples.

Example 1 $a^4 \times a^{-5} = a^{(4 + (-5))} = a^{-1} \left(\text{or } \dfrac{1}{a}\right)$.

Example 2 $x^3 \div x^5 = x^{(3 - 5)} = x^{-2} \left(\text{or } \dfrac{1}{x^2}\right)$.

Example 3 $6x^{-2} \div 2x^{-5} = 3x^{(-2 - (-5))} = 3x^{(-2 + 5)} = 3x^3$.

Practice – Unit 14 Algebraic statements and equations

1 Write the following statements in algebraic form:

(a) I have 25 sweets but then I eat x sweets. How many do I have left?

(b) I am x years old. My brother is 5 years younger than me. How old is he?

(c) John has £y in his pocket. Pete has twice as much as John. How much has Pete got?

(d) If your height is x cm, then what is your height in metres? (1 m $=$ 100 cm)

*2 Write the following in the simplest form possible:

(a) $2a \times 5$.　　(b) $a \times b$.　　　　(c) $a \times b \times -c$.

(d) $20a \div 5$.　　(e) $-45ab \div 9b$.　　(f) $-35a \div -5a$.

*3 Simplify:

(a) $3a + 5a + a$.　　　　　　(g) $2a \times 5a$.

(b) $8a - 3a$.　　　　　　　　(h) $a^4 \times a^5$.

(c) $2a + 7b + 3a + 4b$.　　　(i) $a^6 \div a^2$.

(d) $7a + 2b - 4a - 6b$.　　　(j) $a^2 \times a^3 \times a$.

(e) $a^2 + 2a + 3a^2 - a$.　　(k) $a^2 \times a^{-3}$.

(f) $a \times a \times a$.　　　　　(l) $a^{-2} \div a^{-5}$.

*4 Remove the brackets and simplify:

(a) $2(a + 5)$.　　(b) $3(2a - 5)$.　　(c) $a(2a + 7)$.

(d) $4(a + 3) + 2(3a - 1)$.　　(e) $5(2a - 1) - 3(4a + 1)$.

*5　Factorize: (a) $5a + 10b$.　(b) $3f + 12g$.　(c) $6x^2 - 18x$.

(d) $3a^2bc + 6abc$.

*6 If $p = 3, q = 5$ and $r = 10$, work out the value of the following expressions:

(a) pq.　(b) $4qr$.　(c) $2pq + 3qr$.　(d) $pq + qr + pr$.

(e) $5qr - 2pq$.　(f) pqr.　(g) $4pq \div 2r$.　(h) $\frac{1}{2}pr$.

(i) $p^2 + q^2$.　(j) $2r^2$.

*7 The formula $C = \dfrac{5}{9}(F - 32)$ may be used to change °F to °C.

(a) Find the value of C when $F = 5$.

(b) Find the value of F when $C - 5$.

*8 Solve the following equations:

(a) $5x = 30$.　(b) $3g = 17$.　(c) $x - 7 = 11$.

(d) $x + 5 = 12$.　(e) $2x - 11 = 25$.　(f) $17 - 3x = 2$.

(g) $3x - 1 = 2x + 5$.　(h) $4(m + 3) = 20$.

(i) $\dfrac{y}{4} = 7$.　(j) $\dfrac{3}{4}a = 15$.

*9 For each statement write down and solve an equation to find the numbers:

(a) Trebling a number and then taking away 5 gives the same result as doubling the number and adding 2.

(b) When I add 4 to a number I get the same result as halving the number and adding 10.

*10 (a) If $V = IR$ then write R in terms of V and I.

(b) If $v = u + ft$ then write t in terms of v, u and f.

(c) If $V = \frac{1}{3}r^2h$ then write r in terms of V and h.

(d) If $R = \frac{SL}{a}$ write S in terms of R, L and a.

*11

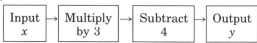

(a) Use the flow diagram to find a formula for y in terms of x.

(b) If you input the number 5, what do you output?

(c) If you output the number 20, what did you input?

(d) Reverse the formula to find x in terms of y.

Unit 15 Coordinates and graphs

Aims of the unit

To revise:

1 The use of coordinates,
2 Constructing tables of values,
3 Drawing and interpreting graphs,
4 Graphs in practical situations,
5 The idea of a gradient.

Coordinates

Coordinates are known as an ORDERED PAIR of numbers since they are always given in the form (x, y), i.e. the x value first, then the y value. We can identify the points marked on the graph (Fig. 1).

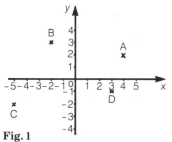

A is the point $(4, 2)$
B is the point $(-2, 3)$
C is the point $(-5, -2)$
D is the point $(3, -1)$

Fig. 1

NB: The point $(0, 0)$ is known as the ORIGIN.

Tables of values

In order to plot a graph of a straight line or curve we must first construct a table of values from the equation of the line or curve.

Example 1 Plot a table of values for the equation

$$y = 2x + 3$$

for x values ranging from $x = -3$ to $x = +3$ often written as $-3 \leqslant x \leqslant 3$ (see Table 1).

Table 1

x	-3	-2	-1	0	1	2	3
$2x$	-6	-4	-2	0	2	4	6
$+3$	$+3$	$+3$	$+3$	$+3$	$+3$	$+3$	$+3$
y	-3	-1	$+1$	$+3$	$+5$	$+7$	$+9$

So, from Table 1, $x = -3, y = -3$, i.e. $(-3, -3)$ and when $x = -2$, $y = -1$, i.e. $(-2, -1)$ and so on.

Example 2 Plot a table of values for the equation

$$y = x^2 - 3x + 2$$

for $-2 \leqslant x \leqslant 4$ (see Table 2).

Table 2

x	-2	-1	0	1	2	3	4
x^2	4	1	0	1	4	9	16
$-3x$	6	3	0	-3	-6	-9	-12
$+2$	2	2	2	2	2	2	2
y	12	6	2	0	0	2	6

So, from Table 2 when $x = -2, y = 12$, i.e. $(-2, 12)$ etc.

Drawing and interpreting graphs

Having constructed the tables of results we are then in a position to plot graphs based on these results.

From Table 1 we can draw the graph of $y = 2x + 3$ (Fig. 2).

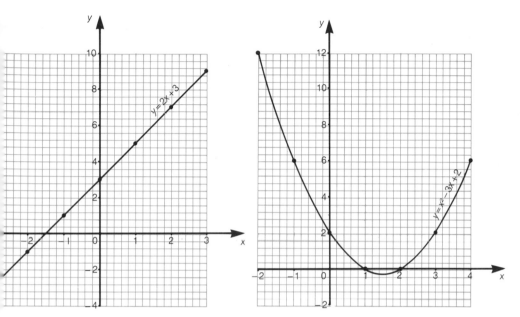

Fig. 2   **Fig. 3**

From Table 2 we can draw the graph of $y = x^2 - 3x + 2$ (Fig. 3).

NB: If you are asked to draw a graph like the ones here, or any other type for that matter, ensure that you use the scale given in the question for the x-axis and the y-axis. Failure to do this will result in a loss of marks.

Graphs in practical situations

Other useful types of graph are the conversion graphs to convert from one currency to another (see Unit 8), and the distance/time graphs shown in Unit 9 from which much information can easily be read, for example finding distances travelled in given times or the time taken to travel given distances.

Gradients

Gradients are either positive or negative depending on which way the line slopes – see Fig. 4.

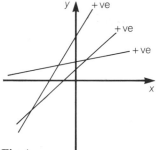

Fig. 4

Positive gradients: lines sloping upwards from left to right have a +ve gradient.

Negative gradients: lines sloping downwards from left to right have a −ve gradient.

We can easily find the gradient of a straight line by dividing the change in *upward* distance by the change in distance *to the right* horizontally.

See the examples in Figs. 5 and 6 of calculating the gradient of a straight line.

One has a positive gradient (Fig. 5), the other a negative gradient (Fig. 6).

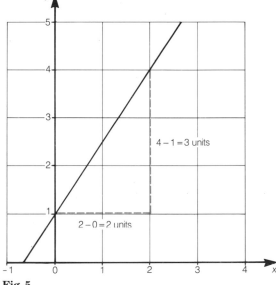

Fig. 5

Gradient of line $= \dfrac{3 \text{ (upwards)}}{2 \text{ (to the right)}}$

Hence the gradient $= \dfrac{3}{2}$ or $1\frac{1}{2}$

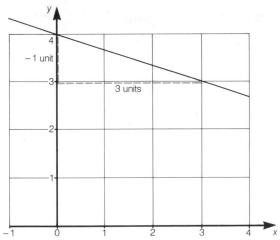

Fig. 6

Gradient of line $= \dfrac{-1}{3} \begin{array}{l} \text{(downwards)} \\ \text{(to the right)} \end{array}$

Hence the gradient $= \dfrac{-1}{3}$

NB: In the case of the distance/time graphs the gradient of the line gives the speed. Fig. 7 shows a typical example.

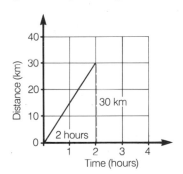

Speed

$= \dfrac{\text{distance travelled}}{\text{time taken}}$

$= \text{gradient of graph}$

In this case,

$\text{speed} = \dfrac{30 \text{ km}}{2 \text{ hours}} = 15 \text{ km/hour}$

Fig. 7

Practice – Unit 15   Coordinates and graphs

1 In Fig. 8,
ABC is an
isosceles
triangle with
A (4, 2)
B (1, 1)
C (1, 3).

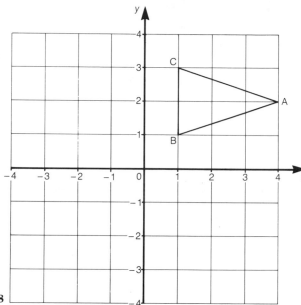

Fig. 8

(a) If the triangle is reflected in the y-axis the image is $A_1 B_1 C_1$.
What are the coordinates of: $A_1($), $B_1($), $C_1($)?
(b) If the original triangle ABC is reflected in the x-axis the image is
$A_2 B_2 C_2$. What are the coordinates of: $A_2($), $B_2($), $C_2($)?
*2   Use the graph of Fig. 9 to find the speeds of:
(a) the car,
(b) the cyclist,
(c) the walker.

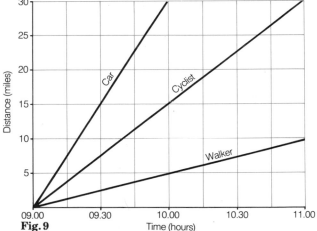

Fig. 9

*3 The table below shows corresponding values of x and y.

x	0	1	2	3	5	10
y	−5	−3	−1	1	5	15

Plot these points on the axis below and join up the points (Fig. 10).

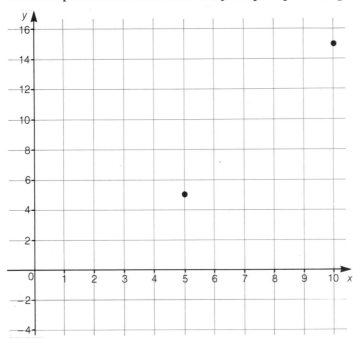

Fig. 10

(a) What is the value of y when $x = 4$?
(b) What is the value of x when $y = 13$?
(c) What is the gradient of the line?

Unit 16 Geometrical shapes and terms

Aims of the unit

To revise:
1 Types of angle,
2 Angles and parallel lines,
3 Triangles and quadrilaterals,
4 Congruence and similarity,
5 Polygons,
6 Simple circle theorems.

Types of angles

We must be able to recognize the different types of angle shown in Fig. 1.

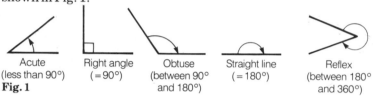

Acute
(less than 90°)
Right angle
(=90°)
Obtuse
(between 90° and 180°)
Straight line
(=180°)
Reflex
(between 180° and 360°)

Fig. 1

NB: Make sure that you know how to measure the size of an angle using a protractor. Remember also that two angles adding up to 90° are called COMPLEMENTARY angles and two angles adding up to 180° are called SUPPLEMENTARY angles (or angles on the same straight line).

Angles and parallel lines

When two lines cross each other (Fig. 2) we must recognize that the VERTICALLY OPPOSITE angles are equal.

α
β
β
α

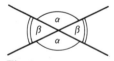

Fig. 2

In Fig. 3 the lines AB and CD are parallel to each other and the straight line EF cuts across both at the points X and Y respectively. This enables us to identify some other types of equal angles.

X
F
A
B
α°
α°
β°
Y
α°
C
D
E

Fig. 3

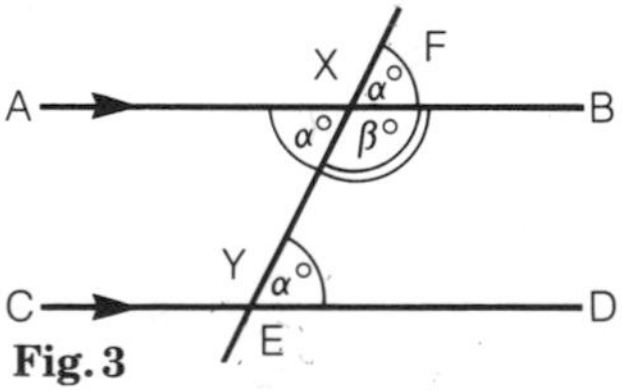

$\angle BXF = \angle DYF = \alpha°$ (known as CORRESPONDING angles)
also
$\angle DYF = \angle AXE = \alpha°$ (known as ALTERNATE angles)

This is confirmed by looking again at Fig. 2.

α° and β° are SUPPLEMENTARY because they total 180° (see above).

Triangles and quadrilaterals

There are 4 different types of triangle that we should be able to recognize (see Fig. 4).

| All sides and angles different — Scalene triangle | All sides and angles equal (60°) — Equilateral triangle | Two sides and two angles equal — Isosceles triangle | One of the angles is a right angle — Right-angled triangle |

Fig. 4

NB: Whatever the type of triangle, the sum of the interior angles always totals 180°.

Similarly there are several different types of quadrilateral (4-sided figures) – see Fig. 5. Some of these we have already encountered in Unit 12 (Area) and Unit 10 (Symmetry).

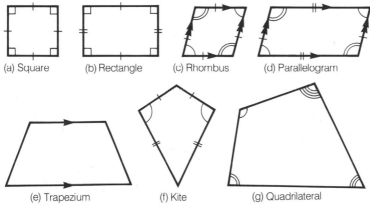

(a) Square   (b) Rectangle   (c) Rhombus   (d) Parallelogram

(e) Trapezium   (f) Kite   (g) Quadrilateral

Fig. 5

Some properties of the above quadrilaterals:

(a) SQUARE – 4 equal sides, 4 right-angles.

(b) RECTANGLE – Opposite sides equal, 4 right-angles.

(c) RHOMBUS – All sides equal, opposite sides parallel, diagonally opposite angles equal.

(d) PARALLELOGRAM – Opposite sides equal and parallel, diagonally opposite angles equal.

(e) TRAPEZIUM – One pair of opposite parallel sides.

(f) KITE – One pair of diagonally opposite equal angles – made up of two isosceles triangles on the same base.

(g) QUADRILATERAL – No sides of the same length and no angles equal.

NB: The sum of the interior angles of any QUADRILATERAL is always 360°.

Congruence and similarity

Figures which have the same size and shape are called CONGRUENT figures. As examples, consider some pairs of congruent triangles.

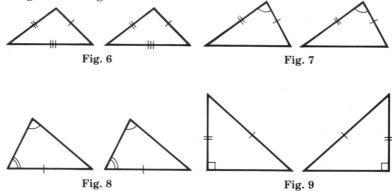

Fig. 6   **Fig. 7**

Fig. 8   **Fig. 9**

In Fig. 6 the pair of triangles have corresponding sides equal in length and are therefore congruent.

In Fig. 7 the pair of triangles have two pairs of corresponding equal side lengths plus the same angle between the two given sides – once again they are congruent.

In Fig. 8 the pair of triangles have two pairs of equal angles plus a corresponding side length equal in both – hence they are congruent.

In Fig. 9 both triangles have a right-angle, the longest side (hypotenuse) of the same length and one other side length equal in each triangle and hence the two triangles are congruent.

If two figures are SIMILAR then corresponding angles are equal and corresponding side lengths are in the same ratio to each other.

Once again, as an example, we will look at a pair of triangles (Fig. 10).

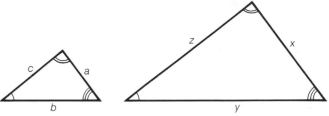

Fig. 10

NB: The ratio of the side lengths will be $\dfrac{a}{x} = \dfrac{b}{y} = \dfrac{c}{z}$

Polygons

This is the general name given to a shape with three or more sides, i.e.:

3 sides – TRIANGLE
4 sides – QUADRILATERAL
5 sides – PENTAGON
6 sides – HEXAGON

7 sides – HEPTAGON
8 sides – OCTAGON
9 sides – NONAGON
10 sides – DECAGON

If the word REGULAR is used with a particular polygon it means that the polygon in question has equal side lengths and equal interior angles.

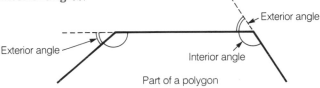

Fig. 11 Exterior angles of a polygon

The sum of the exterior angles of any polygon = 360°.

NB: In the case of a REGULAR polygon all the exterior angles will be equal and hence for a regular octagon (eight sides) each exterior angle will be $\dfrac{360°}{8} = 45°$.

Now, having found the value of an exterior angle we can find the size of an interior angle by subtracting the exterior angle from 180° since: Exterior angle + Interior angle = 180° (they are, of course, SUPPLEMENTARY). Hence, the size of the interior angle of a regular octagon will be 180° − 45° = 135°.

Simple circle theorems

We need to be aware of two properties of a circle.

(a) A diameter of a circle forms •
a right-angle at the
circumference of the circle
(see Fig. 12).

(b) A tangent to a circle is
always at right-angles to the
radius at the point of contact
(see Fig. 13).

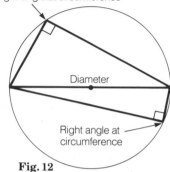

Right angle at circumference

Diameter

Right angle at
circumference

Fig. 12

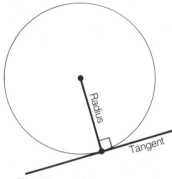

Radius

Tangent

Fig. 13

Practice – Unit 16   Geometrical shapes and terms

1 Work out the angles
denoted by x in Fig. 14.

(a)

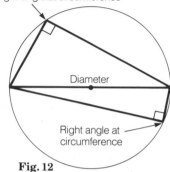

88°

75°

x

33°

(b)

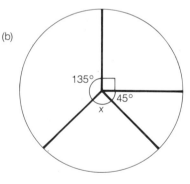

135°

45°

x

(c)

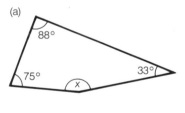

128°

y

x

Fig. 14

*2 In Fig. 15, BCD and ECF are straight lines. Angle ABC = 96°,
angle BCF = 78°, angle CFA = 116° and CE = ED. Find angles
marked x, y and z.

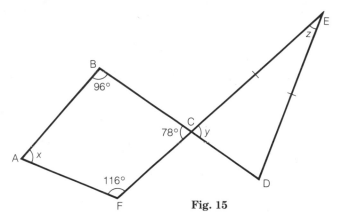

Fig. 15

*3 ABCDEF is a regular hexagon
(Fig. 16).
(a) Find the value of $x°$.
(b) Find the value of $y°$.
(c) How many lines of symmetry
has a regular hexagon?

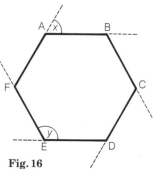

Fig. 16

*4

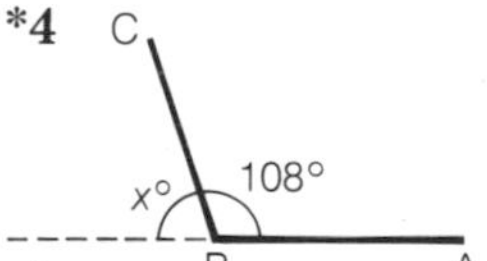

Fig. 17

In Fig. 17 ABC forms part of a
larger regular polygon:
(a) Find the value of $x°$.
(b) How many sides will the
polygon have when complete?
(c) Draw in a sketch of the
completed polygon.
(d) Name the polygon.

*5 (a) The size of the exterior angle of a regular 12-sided
polygon = ...°.
(b) The size of the interior angle of a regular 15-sided polygon = ...°.
(c) If a regular polygon has an interior angle of 144° then the
exterior angle = ...° and the polygon will have ... equal sides.

Unit 17 Scale drawing

Aims of the unit

To revise:
1 Construction of triangles using drawing instruments,
2 Bearings,
3 Reading and making of scale drawings,
4 Enlargements.

Construction of triangles using drawing instruments

For construction questions we obviously need a sharp pencil, a ruler and a pair of compasses.

Example Construct a triangle with side lengths of 8 cm, 6 cm and 4 cm (Fig. 1).

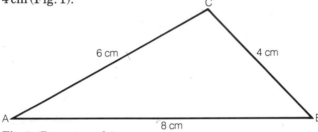

Fig. 1 (Drawn to scale)

Stages of construction:
1 Draw base AB 8 cm long (i.e. longest side).
2 Open up compasses to 6 cm and swing an arc from A.
3 Open up compasses to 4 cm and swing an arc from B.
4 Point C is at the intersection of the two constructed arcs.
5 Join A to C and B to C to complete the triangle.

Once again, you need to be able to measure the angles by correctly using a protractor. In Fig. 1 $\angle A = 30°$, $\angle B = 46°$, $\angle C = 104°$.

Bearings

The directions given by the compass are known as bearings, measured in degrees from 0° to 360°.

They are measured in a clockwise direction from N (north) and are given in 3-figure form, as shown in Fig. 2. Hence a bearing of 5° is actually written 005°. A bearing of 55° is written 055° and a bearing of 165° is written 165° etc.

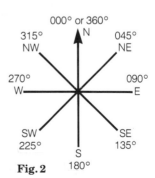

Fig. 2

Example In Fig. 3 find the
bearing of B from A,
then find the bearing of
A from B.

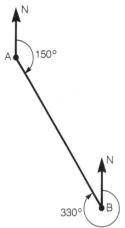

Bearing of B from A = 150°.
Bearing of A from B = 330° etc.

Fig. 3

Scale drawings and enlargements

Obviously it is quite impossible to draw maps and plans to a full scale
size. In order to overcome the problem the map or plan is drawn to a
smaller scale in strict ratio to the full size. The map or plan is thus an
accurate replica of the actual figure. We can of course either scale
things up or scale them down whichever is appropriate. The amount
by which the actual is either enlarged or made smaller is known as
the SCALE FACTOR. If a figure is doubled in size then we say that
the new figure has a scale factor of 2. If, however, the figure is halved
in size then we say that the new figure has a scale factor of $\frac{1}{2}$ etc.

Example 1 The diagram Fig. 4 is an accurate plan of a room in a
house, drawn using a scale of 2 cm to represent 1 m. Use a ruler to
measure in centimetres and then convert into metres.
(a) The length of the room.
(b) The width of the window.
(c) If you were standing at the point P looking through the window,
describe through what angle and in which direction you would need
to turn to face the door.

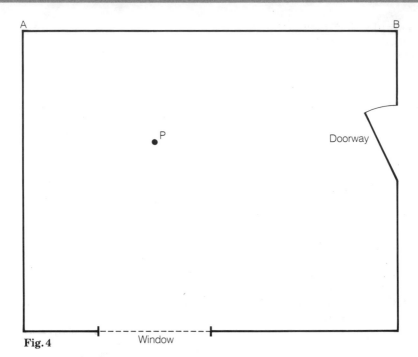

Fig. 4

(a) Length of room = 10 cm on plan = 5 metres actually.
(b) Width of window = 3 cm on plan = $1\frac{1}{2}$ metres actually.
(c) You would need to turn through 90° anticlockwise in order to face the door.

Example 2 In Fig. 5, if AF corresponds to GL draw an enlargement GHIJKL of the original shape ABCDEF.

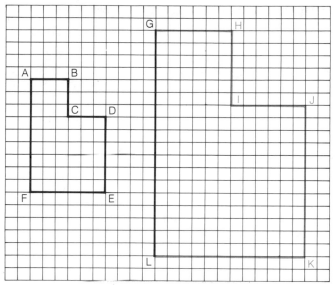

Fig. 5

Practice – Unit 17   Scale drawings and enlargements

1 In the diagram Fig. 6 ABC represents a triangular field.
AB = 50 m, BC = 85 m and angle ABC = 40°.

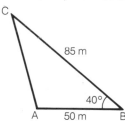

(a) Using a scale of 1 cm to represent 10 m draw an accurate plan of the field.
(b) Find the length of the side AC of the field.
(c) Find the size of angle BAC.

Fig. 6 (Not drawn to scale)

2 The diagram Fig. 7 shows part of an accurately drawn chart.

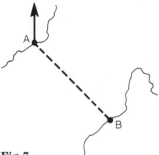

Fig. 7

By measuring, find the bearing of port B from port A.

*3 The diagram Fig. 8 shows three marker buoys used in a yacht race. The bearings of some parts of the route are given.

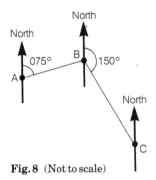

Fig. 8 (Not to scale)

(a) Join AC to show the last part of the race.
(b) Calculate the size of angle ABC.
(c) What is the bearing of A from B?
(d) What is the bearing of B from C?
(e) If the bearing of A from C is 295°, what is the bearing of C from A?

*4 Reduce the shape ABCDE to half size A'B'C'D'E' in Fig. 9.

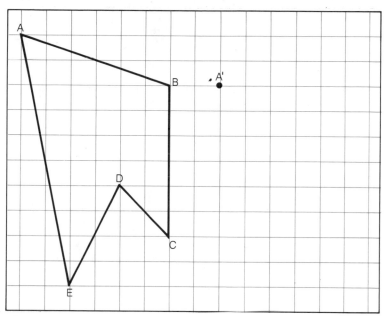

Fig. 9

Unit 18 Transformations and tessellations

Aims of the unit

To revise:
1 Reflections,   3 Translations,
2 Rotations,   4 Tessellations.

A transformation is simply a change. Those that we are concerned with are simple transformations of plane figures.

Reflections

Example Consider Fig. 1.

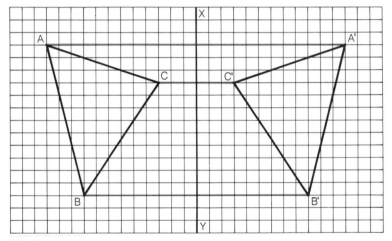

Fig. 1

Triangle ABC, when reflected in the MIRROR, LINE XY, produces an image A′B′C′. Hence, △A′B′C′ is the REFLECTION OF △ABC in the line XY.

It is important to note that
(a) Size and shape are unchanged.
(b) The order of points is reversed.
(c) Lines joining points to their image points are perpendicular to the mirror line XY and are also bisected by the mirror line.
(d) The equation (position) of the mirror line must be given.

Rotations

Example consider Fig. 2. The △XYZ has been rotated through 90° clockwise (−90°) about the CENTRE OF ROTATION, O.

The image of $\triangle XYZ$ is $\triangle X'Y'Z'$. It is important to note the following:

(a) Size and shape are, again unchanged.

(b) The *centre of rotation*, O, must remain static.

(c) The angle between the line joining a point to the centre of rotation and the line joining the corresponding image points to the centre of rotation is called the ANGLE OF ROTATION.

(d) We say that an anticlockwise rotation is positive and a clockwise rotation is negative.

(e) The centre of rotation and the angle and direction of rotation must be given.

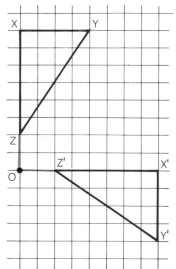

Fig. 2

Translations

A translation is a movement of a figure from one position to another (its image).

Example Consider Fig. 3.

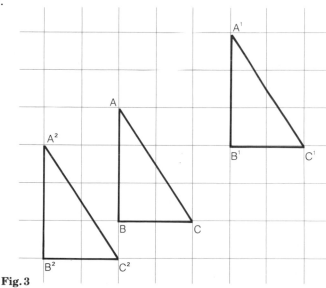

Translation 1: $\triangle ABC$ is moved (translated) to $A^1B^1C^1$ by an 'x' shift of $+3$ and a 'y' shift of $+2$.

Translation 2: $\triangle ABC$ is moved to $A^2B^2C^2$ by an 'x' shift of -2 and a 'y' shift of -1.

It is important to note the following:

Fig. 3

(a) Size and shape are unchanged.
(b) All lines joining points to images are parallel for each translation.
(c) Every point moves.
(d) The 'x' and 'y' shifts need to be given usually in matrix form, known as a COLUMN VECTOR.

In the example above

Translation 1 $= \begin{pmatrix} 3 \\ 2 \end{pmatrix}$ and Translation 2 $= \begin{pmatrix} -2 \\ -1 \end{pmatrix}$

Tessellations

Example 1 Fig. 4 shows that regular hexagons will fit together, forming a flat surface with *no gaps*.

Fig. 4

Shapes that do this are said to TESSELLATE.

NB: Not all regular polygons will tessellate.

Example 2 By drawing appropriate lines, show how the shaded shape can form a tessellation of the whole grid in Fig. 5.

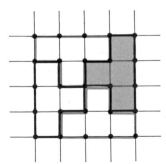

Fig. 5

Practice – Unit 18 Transformations and tessellations

1 Patterns based on rotations, reflections and translations of a
simple shape are used in wallpaper designs.
Complete the pattern in Fig. 6.

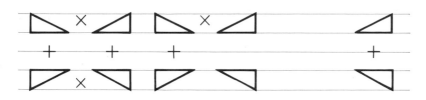

Fig. 6

2 The column vector $\begin{pmatrix} 3 \\ 1 \end{pmatrix}$ means 'move the shape along to the right
3 units – then move the shape up 1 unit'.

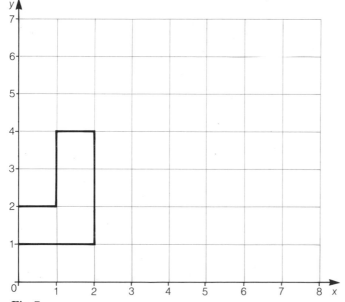

Fig. 7

On Fig. 7, draw the position of the shape after a translation of $\begin{pmatrix} 3 \\ 1 \end{pmatrix}$.

Draw the new position after a further translation of $\begin{pmatrix} 2 \\ 2 \end{pmatrix}$.

3 Show, by drawing five more shapes on the grid of Fig. 8, how the shape tessellates.

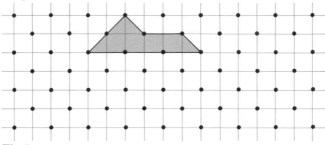

Fig. 8

Unit 19   Pythagoras' Theorem and trigonometry

Aims of the unit

To revise:

1 Pythagoras' Theorem,
2 Sine, cosine and tangent,
3 Angles of elevation and depression,
4 Application of trigonometry to calculate a side length or an angle of a right-angled triangle.

Pythagoras' Theorem

Pythagoras' Theorem states that, for any RIGHT-ANGLED TRIANGLE – the square of the HYPOTENUSE length is exactly equal to the sum of the squares of the lengths of the other two sides.

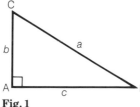

Fig. 1

And so, in Fig. 1, with a being the hypotenuse (longest side, opposite the right-angle) length and b and c being the other two side lengths, Pythagoras' Theorem says that
$$a^2 = b^2 + c^2$$

Example 1 In Fig. 2 find the value of x cm.

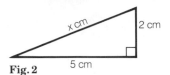

Fig. 2

In this case, applying
Pythagoras' Theorem

$$x^2 = 2^2 + 5^2$$
$$= 4 + 25 = 29$$

x is now found by using the calculator to find the square root of 29.
Hence $\sqrt{29} = 5{\cdot}385$, i.e. $5{\cdot}4$ cm (correct to 1 decimal place).

Example 2 In Fig. 3 find the value of y cm.

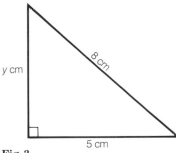

Fig. 3

In this case we are given the
hypotenuse length and so,
applying Pythagoras' Theorem:
$$8^2 = 5^2 + y^2$$
$$64 = 25 + y^2$$

Rearranging, $64 - 25 = y^2$, $\therefore y^2 = 39$
and hence $y = \sqrt{39}$ cm $= 6{\cdot}2$ cm (correct to 1 decimal place).

Trigonometric ratios – sine, cosine and tangent

We are looking here for relationships between side lengths and angle
sizes; but we must remember that these apply, once again, only to
right-angled triangles.

Any right-angled triangle may be labelled as shown in Fig. 4.

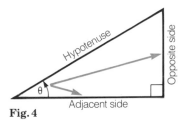
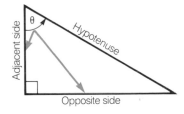

Fig. 4

NB: The ADJACENT SIDE is the side (other than the hypotenuse)
next to the angle $\theta°$ and the OPPOSITE SIDE is the side opposite the
angle $\theta°$.

There are three general relationships between sides and angles in
a right-angled triangle. They are known as the trigonometric (trig.)
ratios. The three ratios are as follows:

$$\text{tangent of } \theta° \ (\tan \theta) = \frac{\text{opposite side length}}{\text{adjacent side length}}$$

$$\text{sine of } \theta° \ (\sin \theta) = \frac{\text{opposite side length}}{\text{hypotenuse length}}$$

$$\text{cosine of } \theta° \ (\cos \theta) = \frac{\text{adjacent side length}}{\text{hypotenuse length}}$$

There are obviously two different types of problem possible:

(a) Given a side length and an angle (other than the right-angle) we can find a second side length.

(b) Given 2 side lengths we can find an angle (other than the right-angle).

Following are a couple of typical examples illustrating points (a) and (b) above.

Example 1 In Fig. 5 find the lengths of AB and BC.

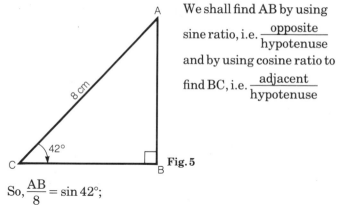

We shall find AB by using

sine ratio, i.e. $\dfrac{\text{opposite}}{\text{hypotenuse}}$

and by using cosine ratio to

find BC, i.e. $\dfrac{\text{adjacent}}{\text{hypotenuse}}$

Fig. 5

So, $\dfrac{\text{AB}}{8} = \sin 42°$;

rearranging gives AB = 8 × sin 42°
using the calculator AB = 5·353 044 9
and so AB = 5·35 cm (to 3 sig. figs.).

NB: Make sure that you know how to use your calculator to produce this answer.

Then, $\dfrac{\text{BC}}{8} = \cos 42°$;

rearranging gives BC = 8 × cos 42°
using the calculator BC = 5·945 158 6
and so BC = 5·95 cm (to 3 sig. figs.).

Example 2 In Fig. 6 calculate the value of $\theta°$.

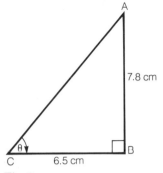

AB is the opposite side length
BC is the adjacent side length
Hence the correct ratio required
will be tangent.

Hence $\tan \theta = \dfrac{\text{opposite}}{\text{adjacent}}$

$\qquad = \dfrac{7 \cdot 8}{6 \cdot 5}$

i.e. $\tan \theta = 1 \cdot 2$

$\qquad \theta° = \tan^{-1} 1 \cdot 2$

Fig. 6

NB: At this stage be sure that you understand how to operate your
'2nd function' key for $\tan^{-1}$.

$\theta = 50 \cdot 194\ 429°$, which to the nearest tenth of a degree will be
$50 \cdot 2°$; hence, $\theta° = 50 \cdot 2°$.

Angles of elevation and depression

The angle of elevation is the angle between the horizontal and the
object (when the object is above the observer) (Fig. 7a).

The angle of depression is the angle between the horizontal and
the object (when the object is below the observer) (Fig. 7b).

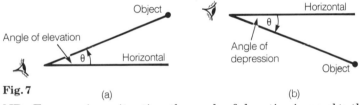

Fig. 7   (a)   (b)

NB: For any given situation, the angle of elevation is equal to the
angle of depression (they are alternate angles). See Fig. 8.

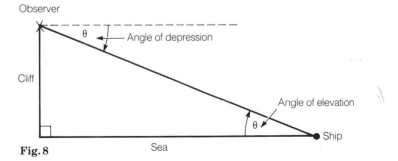

Fig. 8

Problems involving trigonometry ratios

We can now apply our trig. knowledge to specific problems such as the angle a ladder makes with a wall, or the height of a tree, or even the distance from the shore to a boat out at sea.

Example 1 How far up a wall will a ladder 5 m long reach if the ladder makes an angle of 38° with the ground?

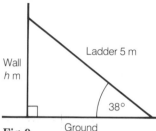

Fig. 9

See Fig. 9. Let the height the ladder reaches up the wall be h m.

Then $\dfrac{h}{5} = \sin 38°$

$\left(\text{i.e. } \dfrac{\text{opposite}}{\text{hypotenuse}}\right)$

$h = 5 \times \sin 38° = 3\cdot08 \text{ m}$

Hence, the ladder will reach 3·08 m up the wall.

Example 2 Find the angle of elevation of the top of a building 20 m high from a point on the ground 15 m away from the building.

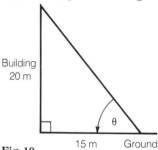

Fig. 10

See Fig. 10. Let the angle of elevation be $\theta°$.

Then $\dfrac{20}{15} = \tan \theta$

$1\cdot333 = \tan \theta$

Hence $\theta° = 53\cdot1°$ (to the nearest tenth of a degree).

So the angle of elevation of the top of the building from the point on the ground is 53·1°.

Example 3 A ship sails from a point A to a point C, a distance of 80 km, on a bearing of 050°. Point B is due north of A and due west of C. Find the distance from A to B.

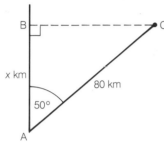

See Fig. 11. Let the distance AB be x km.

Then $\dfrac{x}{80} = \cos 50°$

$\therefore x = 80 \times \cos 50° = 51·4\,\text{km}$
So, B is 51·4 km due north of A.

Fig. 11

Practice – Unit 19 Pythagoras' Theorem and trigonometry

*1

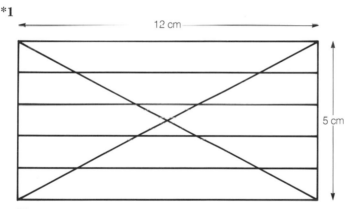

Fig. 12

A model gate is constructed of 6 horizontal bars, 2 vertical bars and 2 diagonals in the shape of a rectangle measuring 5 cm by 12 cm (Fig. 12). The model is made of plastic tubing. Calculate:
(a) The length of one diagonal
(b) The total length of plastic tubing needed to make the model.

*2

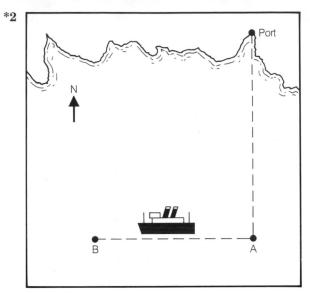

Fig. 13

In Fig. 13, a ship sails from port for 12 km due south to point A. It then changes course to due west and sails for a further 10 km to point B. How far is the ship at point B from the port?

*3 The angle of elevation of the top of a tree (Fig. 14) is 22° taken from a point on the horizontal ground 50 m away from the base of the tree. How tall is the tree?

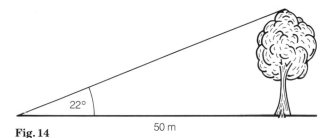

Fig. 14   50 m

The following questions are suggested for practice purposes and are of Foundation/Intermediate Level standard:

1 A dice is thrown with its 6 faces numbered 1, 2, 4, 5, 6, 8. What is the probability of the top face showing an odd number?

2

GRAND SALE
40% OFF
all marked prices

Fig. 1

(a) Mr Singh bought a wheelbarrow. Its marked price was £18.00
 (i) How much was taken off this marked price?
(ii) How much did Mr Singh pay for the wheelbarrow?

(b) The following day the shop sold the same wheelbarrows at half the marked price. How much more would Mr Singh have saved by waiting another day?

3

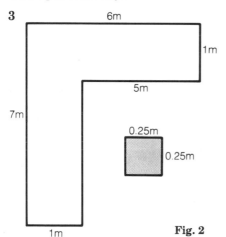

Fig. 2

(a) The diagram is of a corridor 1 m wide. Work out the area of the floor of the corridor?

(b) The corridor floor is covered with square carpet tiles with sides of 0.25 m as shown. How many carpet tiles are needed to cover 1 square metre of floor?

(c) How many tiles will be needed to cover the corridor?

4 A 30 cm ruler is cut into 2 pieces 12 cm and 18 cm long. The piece 12 cm long weighs 10 grams. What is the weight of the 18 cm portion?

5 Jack and Jill are playing a numbers game. Jill has chosen a formula and Jack has to guess what it is. He gives Jill a number and she tells him the result after using her formula.
Here are the results for Jack's first four numbers:

Jack's number	Jill's result
4	13
7	22
3	10
12	37

(a) If Jack's next number is 9, what do you think will be Jill's result?

(b) Explain how Jill obtains her results.

6

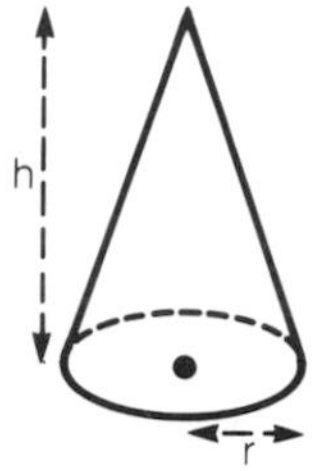

Fig. 3

The volume of a cone is approximately found using the formula V = $1.05r^2h$.

Calculate the volume of a cone when r = 6 cm and h = 10 cm (Answer in cm³.)

7 Mary has three envelopes to post. The post office is closed but she has a supply of 20p and 15p stamps. She wants to waste as little money as possible. Work out which stamps she should put on an envelope if the required postage is:

(a) 34p,   (b) 65p,   (c) 69p.

8 The temperatures at midday in Moscow on the seven days of one week were as follows:

Sunday	−3°C	Thursday	4°C
Monday	−2°C	Friday	2°C
Tuesday	−8°C	Saturday	−7°C
Wednesday	0°C		

(a) Calculate the average (mean) temperature for the week.

(b) On which days was the midday temperature below the average for the week?

9

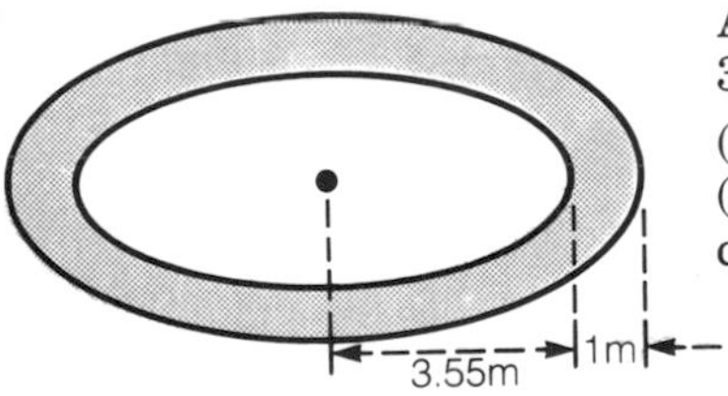

Fig. 4

A circular pond has a radius of 3.55 m.

(a) Taking π to be 3.14, calculate (correct to 2 sig. figs.) the circumference of the pool.

(b) The area of the pond.

(c) The pond is surrounded by a concrete path 1 m wide. Calculate the area of the surface of the path (correct to 2 sig. figs).

10 When Mr Smart was on holiday in Spain, the exchange rate was 196.80 pesetas for £1. He bought an English newspaper which normally cost 25p at home.

(a) How many pesetas were equivalent to 25p?

(b) How many pence were equivalent to 150 pesetas (to the nearest whole number)?

(c) Mr Smart complained that the newspaper cost more than three times as much in Spain. Was he correct? Give a reason for your answer.

11 A man died and left £12,000 in his will to be divided amongst three people in the ratio 3 : 5 : 7. How much did each receive?

12 (a) Find the value of: (i) $2^2 - 5^2$, (ii) $3^5 - 5^3$

(b) Put these three numbers in order of size, smallest first.
$$8.12 \times 10^5, \ 1.72 \times 10^6, \ 2.89 \times 10^5$$

13

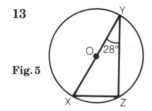

Fig. 5

A circle has a diameter XY. Z is a point on its circumference. Angle XYZ is 28°. Calculate angle XYZ (giving reasons).

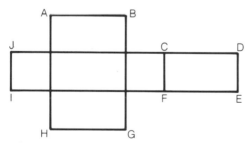

Fig. 6

The diagram shows a net consisting of 2 squares and 4 rectangles. It is cut out and folded to make a solid figure.

(a) What is the name given to the solid formed?

(b) Which points of the net meet at A?

15 The diagram shows two lighthouses X and Y, with X due north of Y. From a ship A the bearing of X is 300° and the bearing of Y is 210°

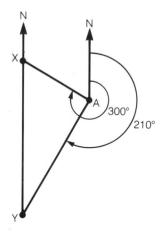

(a) Calculate:
 (i) angle XAY
(ii) angle XYA

(b) The distance XY is 3000 m
Calculate the distances:
 (i) XA
(ii) YA

Fig. 7

Unit 1 Number

1 (a) 25,36 (b) 21,28 (c) 21,34 (d) 125,216
2 (a) 12 (b) 4 (c) 10 (d) 512 (e) 169 (f) 32
3 (a) 16 (b) 7 (c) 24 (d) 105 (e) 90 (f) 15
4 (a) 12 (b) 6 (c) 26 (d) 32 (e) 13 (f) 3
5

	Arsenal	Man. Utd	Leeds	Orient	Total
Week 1	21 461	39 586	31 429	9628	102 104
Week 2	24 072	42 721	27 333	5406	99 532
Week 3	27 624	41 027	29 281	9672	107 604
Week 4	20 467	49 678	26 621	8241	105 007
Club Total	93 624	173 012	114 664	32 947	414 247

6 (a) 1,5 (b) 12 (c) 24 (d) 24 (e) 120 seconds
7 (a) $\frac{1}{5}$ (b) $\frac{1}{2}$ (c) $\frac{1}{16}$ (d) $\frac{1}{8}$ (e) 1
   (f) 2 (g) 5 (h) 2 (i) 3 (j) 10

Unit 2 Fractions

1 (a) $\frac{1}{2}$ (b) $\frac{1}{3}$ (c) $\frac{1}{4}$
2 (a)   (b)

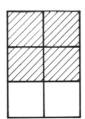
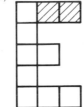
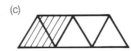

(c)

3 (a) $1\frac{1}{4}$ (b) $\frac{1}{10}$ (c) $11\frac{1}{4}$ (d) $\frac{1}{10}$
   (e) $\frac{1}{2}$ (f) $2\frac{2}{9}$ (g) 4 (h) $4\frac{1}{2}$
4 (a) £3000 (b) £1875 (c) $\frac{7}{20}$
5 (a) 15 000 litres (b) $\frac{1}{3}$ full (c) $\frac{2}{3}$ (d) 10 000 litres
6 (a) £12 (b) £96 (c) $\frac{3}{8}$
7 (a) 50 (b) $\frac{2}{5}$ (c) $\frac{1}{5}$ (d) $\frac{3}{10}$

Unit 2 Decimals

1 $1\frac{1}{2}$ $1\frac{3}{5}$ 1·65   1·66 $1\frac{2}{3}$

2 (a) 7·2   (b) 1·25   (c) 1·44

3 (a) 5·4 pints   (b) 37·5 cm   (c) 6 inches

4 (a) 231 miles   (b) £10.80   (c) 8 gallons   (d) 11 gallons

5 (a) 187·24 secs   (b) 3 mins 7·24 secs   (c) 2·61 secs

6 (a) $6·7 \times 10^{3}$ (b) $8·45 \times 10^{5}$ (c) $7·4 \times 10^{-2}$ (d) $3·8 \times 10^{-4}$

7 (a) 1700   (b) 104 000   (c) 23   (d) 0·62

Unit 2 Percentages

1

Fraction	Decimal	Percentage
$\frac{1}{2}$	0·5	50%
$\frac{1}{4}$	0·25	25%
$\frac{3}{4}$	0·75	75%
$\frac{3}{10}$	0·30	30%
$\frac{2}{5}$	0·4	40%
$\frac{4}{5}$	0·8	80%
$\frac{3}{5}$	0·6	60%
$\frac{7}{25}$	0·28	28%
$\frac{3}{20}$	0·15	15%
$\frac{11}{20}$	0·55	55%
$\frac{2}{3}$	0·666 66	$66\frac{2}{3}\%$
$\frac{1}{3}$	0·333 33	$33\frac{1}{3}\%$
$\frac{1}{8}$	0·125	$12\frac{1}{2}\%$

2 Maths 78%   History 70%   English 65%
Science $\left.\begin{array}{c}\\\\\end{array}\right\}$ 60%
Geography

3 (a) 60% girls   (b) 12 boys
4 (a) £35   (b) £31.50   (c) £6.50   (d) 26%
5 (a) £86.10   (b) £80
6 £900
7 (a) 5%   (b) 150 cm

Unit 3 Directed number

1 (a) >   (b) >   (c) <   (d) >   (e) <
 (f) <   (g) >   (h) >   (i) <   (j) >
2 32°C   **3** 15 m
4 (a) 11 pts   (b) 2 pts   (c) −7 pts   (d) −16 pts
5 Mon   rise 3°C
 Tues   rise 8°C
 Wed   fall 14°C
 Thurs fall 3°C
 Fri   rise 13°C
 Sat   Temperature was −4°C
6 (a) +9   (b) −21   (c) +1   (d) $+\frac{1}{2}$

Unit 4 Ratio proportion and scale

1 (a) 1:4   (b) 1:3   (c) 4:5   (d) 3:4:6
2 (a) 1:4   (b) 12:1   (c) 7:10   (d) 1:5
3 (a) £96:£64   (b) 24 cm:36 cm:60 cm
4 £4.80
5 £121   **6** (a) 400 g   (b) 36 g
7 (a) 12 hours   (b) 4 hours
8 (a) 104 m   (b) 50 m   (c) 162 m
9 (a) 4 km   (b) 10 cm   (c) 8 cm^2
10 Sachet 4p for 5 ml
  Small bottle 3.25p for 5 ml
  Large bottle 3p for 5 ml
  Large bottle is best value

Unit 5 Approximation and estimation

1 (a) TV £160   Calculator £10   Watch £20
   (b) TV £160   Calculator £13   Watch £24
2 25·3 secs   (b) 10·1 secs   (c) 13·7 metres
3 (a) 45° (b) 3 cm (c) 4 m (d) 3 m
4 (a) 50 000 (b) 55 000 (c) 55 000 (d) 54 980
5 (a) 2408·75 (b) 2408·7 (c) 2410 (d) 2400 (e) 2409
6 (a) 10 cm (b) 20 cm (c) 15 cm

Unit 6 Calculators

1

Sum	Estimate	Calculator
28×11	$30 \times 10 = 300$	308
42×97	$40 \times 100 = 4000$	4074
103×78	$100 \times 80 = 8000$	8034
61×121	$60 \times 120 = 7200$	7381

2 You always finish with the number you started with.

3 (a)

13
234 312
18 432 24

(b)

15
270 300
18 360 20

(c)

1 3 3
5 27
5 45 9

4 (a) 55 tins (b) 120 tins (c) 20 layers (d) 20 tins
5 (a)

Sum	Guess	Calculator
$\sqrt{10}$	3·1	3·162 277 7
$\sqrt{20}$	4·5	4·472 136
$\sqrt{50}$	7·1	7·071 067 8
$\sqrt{100}$	10	10

Unit 7 Metric units

1 (a) 500 cm (b) 3000 m (c) 80 mm (d) 4250 g (e) 2750 ml
2 (a) 115 cm (b) 3 times
3 (a) 5 posts (b) 50 cm left over
4 (a) 250 (b) 8 m
5 (a) 0·65 l (b) 2·4 kg (c) 4·2 cm (d) 0·25 m (e) 0·6 km
6 12·78 kg **7** 650 g
8 (a) 80 km (b) 134·4 km (c) 35·2 km

Unit 8 Foreign currency and holidays

1 (a) 890 pesetas (b) 16 dollars (c) £52.87 (d) £26.25
(e) 6305 pesetas
2 £21.88 **3** £4.58 **4** At the duty-free shop; cheaper by £13.33
5 (a) £15 (b) £110 (c) £765 (d) 16 875 pesetas
6 (a) 1380 francs (b) 441 marks (c) £48
7 (a) approx. £19 (b) approx. $21 (c) £5

Unit 8 Hire purchase, discounts and VAT

1 Bed £177 TV £320 Hi-fi £227.50
2 (a) £34 (b) £28.80
3 (a) £20.70 (b) £5.18
4 Labour   £64.00
  Parts   £24.00
  _____
  Total   £88.00
  VAT   £13.20
  _____
  Total   £101.20

5 (a) £5 (b) £5.75
6 (a) £3750 (b) £4115
7 (a) £198 (b) £462 (c) £531.30 (d) £44.28

Unit 8   Simple and compound interest

1 (a) £96   (b) £480
2 (a) £105.60   (b) £98.10   (c) Matthew has more
3 £8000   **4** 5 years
5 £3194.40   **6** (a) Simple interest at $12\frac{1}{2}\%$ (b) £17.60

Unit 8   Household bills

1 (a) 8342   (b) 4668   (c) 3525
2 (a) 111 units   (b) 472 units   (c) 1895 units

3 1819 @ 6p	= £109.14	**4** 216 @ 6.5p	= £14.04
Standing charge	£6.80	Standing charge	£6.80
Total	£115.94	Total	£20.84

5 68 @ 38p	= £25.84
Standing charge	£10.60
Total	£36.44

6 (a) 5729 therms   (b) £473.52   (c) No   (d) £178.18 out
7 (a) 7342   (b) 7762   (c) 441   (d) £113.44
8 (a) 1107   (b) £75.10   (c) £86.37
9 (a) £50.75   (b) £28.30   (c) 45 therms
10 (a) £136.20   (b) £68.10   (c) £6090

Unit 8   Wages

1 £140   **2** £4.80   **3** (a) £5.10   (b) £136   (c) £161.50
4 (a) M: £15; Tu: £21; W: £22.80; Th: £27.75 Fr: £14.70   (b) £101.25
5 (a) £1050   (b) £5670   **6** £90
7 (a) £140   (b) £166.25   (c) 2150 hours   (d) 46 hours
8 (a) £4250   (b) £21 250   (c) £2950   (d) £1950   (e) £162.50

Unit 8 Profit and loss

1 (a) £2 (b) 40%
2 (a) 6p (b) £12 (c) 60%
3 £46 **4** £56
5 £95 **6** (a) £255 (b) 20% (c) 250
7 (a) £46.80 (b) £40.95 (c) £4.95

Unit 9 Speed, distance and time

1 10.15 p.m.
2 (a) 2000 hours (b) 2130 hours (c) 0445 hours (d) 0015 hours
(e) 1210 hours
3 (a) 75 km (b) 20 miles (c) 4 hours (d) 5 hours (e) 5 km/
hour (f) 400 km/hour
4 (a) 172 miles (b) 110 miles (c) 3 hours (d) 40 mph
(e) 40 mph
5 (a) 31 mins (b) 43 mins (c) 1308 (d) 1401 (e) 3 mins
6 (a) 25 miles (b) 71 mph (c) $1\frac{1}{2}$ hours (d) 236 miles
(e) 4 hours (f) 59 mph
7 (a) 5 mins (b) After (c) 8 mph
8 (a) 3 times (b) 30 miles (c) 40 miles (d) $\frac{1}{2}$ hour
(e) 60 mph (f) $\frac{1}{2}$ hour

Unit 10 Symmetry

1 (a) M A T Ⓗ S (b) L Ⓘ O N   **4**
2 (a) Ⓓ I Ⓥ I Ⓓ Ⓔ
(b) Ⓐ Ⓓ Ⓓ I Ⓣ I O N
3 (a) Ⓢ Ⓘ Ⓧ T E E Ⓝ
(b) Ⓢ Ⓗ A R E

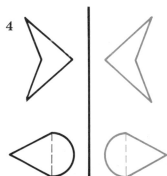

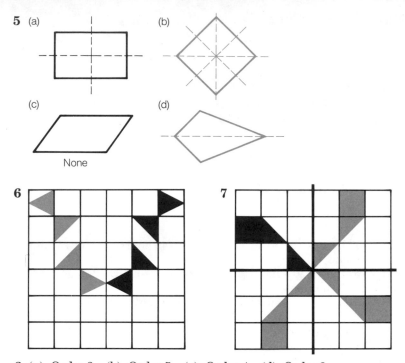

5 (a)

(b)

(c)

None

(d)

8 (a) Order 3   (b) Order 5   (c) Order 4   (d) Order 3
(e) Rectangle; parallelogram

Unit 11 Nets

1 (a)

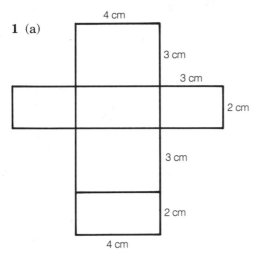

(b) 52 cm² (c) 24 cm³
2 (a) Triangular prism (b) 5 faces (c) 9 edges
3 (a) 3300 cm² (b) 70 cm by 70 cm (c) 1600 cm²

Unit 12 Area, perimeter and volume

1 (a) 16 cm (b) 4 cm (c) 1 cm² **2** a, c, d
3 (a) 150 cm² (b) 27 cm² (c) 123 cm²
4 (a) 1200 m² (b) 28 m (c) 116 m² (d) 6 m²
5 (a) 1200 cm² (b) 120 cm (c) 12 m
6 (a) 36 m (b) £112 (c) 64 m²
7 (a) 1500 cm³ (b) 12 cm³ (c) 125 boxes
8 (a) 50 cm² (b) 48 cm²
9 (a) $1178\frac{1}{4}$ cm³ (b) 471·3 cm² (c) 628·4 cm²
10 (a) 140·18 cm² (b) 5607·25 cm³ (c) 56·07 kg

Unit 13   Tally charts

1

Goals	Tally	Freq.
0	I	1
1	II	2
2	III	3
3	IIII I	5
4		0
5	III	3
6	IIII I II	7
7	IIII	4
8	III	3
9	II	2

2

Class interval	Tally	Freq.
50–54	II	2
55–59	IIII	4
60–64	IIII I IIII	9
65–69	IIII I IIII I III	13
70–74	IIII I IIII I I	11
75–79	IIII I I	6
80–84	III	3
85–89	II	2

Unit 13   Pictograms, bar charts, pie graphs and histograms

1 (a) £30   (b) £80   (c) £300   **2** (a) BL−40   Ford−16
Nissan−5   Vauxhall−20   Peugeot−9   (b) $\frac{1}{18}$ (c) 10%
3 (a) Cheshire   (b) Wensleydale   (c) 7 packs   (d) 49 packs
4 (a) 9 girls   (b) 59 cards   (c) Nil   (d) 3 girls−10%

Unit 13   Means, medians and mode

1 (a) Mean = 1·8   (b) Median = 2   (c) Mode = 2

2

No. of words	1	2	3	4	5	6	7	8	9	10	11	12	13
Freq.	4	2	1	0	0	1	4	3	4	6	3	10	4

(a) Mean = 9 words
(b) Median = 10 words
(c) Mode = 12 words

Unit 13 Simple probability

1 (a) 0 (b) $\frac{3}{10}$ (c) $\frac{3}{10}$ (d) $\frac{1}{5}$ (e) $\frac{7}{10}$
2 (a) $\frac{1}{5}$ (b) $\frac{3}{10}$ (c) $\frac{1}{2}$ (d) Nil
3 $P(5) = \frac{1}{6}$ $P(\text{not } 5) = \frac{5}{6}$
4 (a) $\frac{2}{5}$ (b) $\frac{7}{10}$ (c) $\frac{1}{10}$ (d) $\frac{4}{5}$

Unit 13 Sample spaces and probability trees

1

Tail	(T,1) (T,2) (T,3) (T,4) (T,5) (T,6)
Head	(H,1) (H,2) (H,3) (H,4) (H,5) (H,6)
	1 2 3 4 5 6

(a) $\frac{1}{12}$ (b) $\frac{1}{12}$ (c) Nil

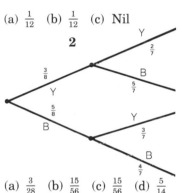

2

(a) $\frac{3}{28}$ (b) $\frac{15}{56}$ (c) $\frac{15}{56}$ (d) $\frac{5}{14}$

Unit 14 Algebraic statements and equations

1 (a) $25 - x$ (b) $x - 5$ (c) $£2y$ (d) $\dfrac{x}{100}\,\text{m}$
2 (a) $10a$ (b) ab (c) $-abc$ (d) $4a$ (e) $-5a$ (f) 7
3 (a) $9a$ (b) $5a$ (c) $5a + 11b$ (d) $3a - 4b$ (e) $4a^2 + a$ (f) a^3
   (g) $10a^2$ (h) a^9 (i) a^4 (j) a^6 (k) a^{-1} (l) a^3
4 (a) $2a + 10$ (b) $6a - 15$ (c) $2a^2 + 7a$ (d) $10a + 10$ (e) $-2a - 8$

5 (a) $5(a+2b)$ (b) $3(f+4g)$ (c) $6x(x-3)$ (d) $3abc(a+2)$
6 (a) 15 (b) 200 (c) 180 (d) 95 (e) 220 (f) 150
(g) 3 (h) 15 (i) 34 (j) 200
7 (a) $-15°C$ (b) $41°F$
8 (a) 6 (b) $5\frac{2}{3}$ (c) 18 (d) 7 (e) 18
(f) 5 (g) 6 (h) 2 (i) 28 (j) 20
9 (a) $3x-5=2x+2$ (b) $x+4=\frac{1}{2}x+10$
　　　$x=7$ 　　　　　　　$x=12$
10 (a) $R=\dfrac{V}{I}$ (b) $t=\dfrac{v-u}{f}$ (c) $r=\sqrt{\dfrac{3v}{h}}$ (d) $S=\dfrac{Ra}{L}$

11 (a) $3x-4=y$ (b) 11 (c) 8 (d) $x=\dfrac{y+4}{3}$

Unit 15　Coordinates and graphs

1 (a) $(-4,2)(-1,1)(-1,3)$ (b) $(4,-2)(1,-1)(1,-3)$
2 (a) 30 mph (b) 15 mph (c) 5 mph
3

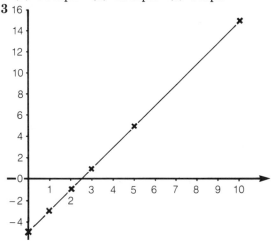

(a) $y=3$
(b) $x=9$
(c) Gradient $=2$

Unit 16 Geometrical shapes and terms

1 (a) 164°   (b) 90°   (c) $x = 52°; y = 128°$
2 $x = 70°$ $y = 78°$ $z = 24°$
3 (a) 60°   (b) 120°   (c) 6   **4** (a) 72°   (b) 5   (d) Pentagon
5 (a) 30°   (b) 156°   (c) 36°; 10

Unit 17 Scale drawings and enlargements

1 (b) 57ʻm   (c) 105°   **2** 134°
3 (b) 105°   (c) 255°   (d) 330°   (e) 115°
4 Angles of same size but all side lengths halved.

Unit 18 Transformations and tessellations

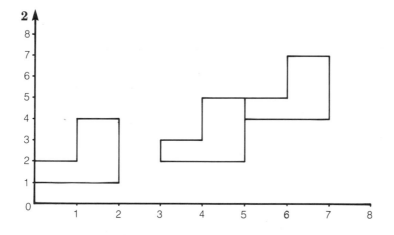

3

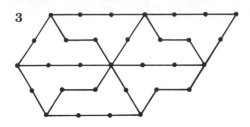

Unit 19 Pythagoras' Theorem and trigonometry

1 (a) 13 cm   (b) 108 cm   **2** 15·62 km   **3** 20·2 m

Unit 20 Sample GCSE Questions

1 $\frac{2}{6}$ or $\frac{1}{3}$ **2** (a) (i) £7.20   (ii) £10.80   (b) £1.80

3 (a) 12 m^2 (b) 16 tiles   (c) 192 tiles

4 15 grams   **5** (a) 28   (b) (Jack's number $\times$ 3) + 1

6 378 cm^3

7 (a) 20p + 15p   (b) $(3 \times 15p) + (1 \times 20p)$ (c) $(2 \times 15p) + (2 \times 20p)$

8 (a) $-2°C$ (b) Sun, Tues, Sat.

9 (a) 22 m   (b) 40 m^2 (c) 25 m^2

10 (a) 49.2 pesetas   (b) 76p   (c) Yes, since $3 \times 25p = 75p$

11 £2400 : £4000 : £5600

12 (a) (i) 7   (ii) 118   (b) 2.89×10^5; 8.12×10^5; 1.72×10^6

13 $<YXZ = 62°$ since $<XYZ = 90°$ (subtended by diameter XY at Z)
or (angle in a semi-circle)

14 (a) Cuboid (box)   (b) J and D

15 (a) (i) $<XAY = 90°$ (ii) $<XYA = 30°$
(b) (i) XA = 1500 m   (ii) YA = 2598 m

Having followed a planned revision programme, you should be well prepared for the examinations. You should be confident about your chances of success.

On some papers you may meet different kinds of questions.

1 Short answer questions The answer may be easily obtainable, but always show the examiner how you have arrived at the answer. You may also be expected to complete a table of results – again, where necessary, show your working.

2 Structured questions These are a series of questions about the same situation. Often the questions start easily but get more difficult. It is absolutely vital that you show, in a logical manner, the way in which you have achieved your answers. The reason for this is that the examiner may be able to award Method Marks even if your answers are incorrect. Finally, always make a note to the effect that you have used a calculator in order to achieve your answer.

The following advice may help you achieve your best on the day.

1 Make sure you have a good night's sleep before the examination. Do not become over-tired doing last-minute revision.

2 Make sure you have all the equipment you need ready the night before – pen, pencil, pencil sharpener, rubber, ruler, protractor, compasses and a calculator with a good battery (if required). You may also need a watch, in case you can't see the examination room clock.

3 Arrive in good time for the examination.

4 Make sure you know the length of the paper and number of questions you must attempt.

5 Do not waste time trying to answer questions you cannot do. Leave out these questions and come back to them.

6 Read the questions thoroughly. GCSE questions contain a great deal of information but candidates do not always use it.

7 Write clearly for the benefit of the examiner and cross out neatly (one diagonal line only) any errors you have made.

8 The number of marks for each question is usually given. This should give you some idea of what is required by the examiner.

9 If you finish the paper early, go back and check your answers.